T0419591

HORIZONS IN WORLD PHYSICS

VOLUME 302

HORIZONS IN WORLD PHYSICS

Additional books and e-books in this series can be found on Nova's website under the Series tab.

HORIZONS IN WORLD PHYSICS

HORIZONS IN WORLD PHYSICS

VOLUME 302

ALBERT REIMER
EDITOR

NOTICE TO THE READER

The Publisher has taken reasonable care in the preparation of this book, but makes no expressed or implied warranty of any kind and assumes no responsibility for any errors or omissions. No liability is assumed for incidental or consequential damages in connection with or arising out of information contained in this book. The Publisher shall not be liable for any special, consequential, or exemplary damages resulting, in whole or in part, from the readers' use of, or reliance upon, this material. Any parts of this book based on government reports are so indicated and copyright is claimed for those parts to the extent applicable to compilations of such works.

Independent verification should be sought for any data, advice or recommendations contained in this book. In addition, no responsibility is assumed by the Publisher for any injury and/or damage to persons or property arising from any methods, products, instructions, ideas or otherwise contained in this publication.

This publication is designed to provide accurate and authoritative information with regard to the subject matter covered herein. It is sold with the clear understanding that the Publisher is not engaged in rendering legal or any other professional services. If legal or any other expert assistance is required, the services of a competent person should be sought. FROM A DECLARATION OF PARTICIPANTS JOINTLY ADOPTED BY A COMMITTEE OF THE AMERICAN BAR ASSOCIATION AND A COMMITTEE OF PUBLISHERS.

Additional color graphics may be available in the e-book version of this book.

Library of Congress Cataloging-in-Publication Data

ISBN: 978-1-53617-180-8
ISSN: 2159-2004

Published by Nova Science Publishers, Inc. † New York

CONTENTS

PREFACE

Horizons in World Physics. Volume 302 considers the explanation of (bio)corrosion and (bio)leaching on the base of changes of electric potentials. (Bio)corrosion and (bio)leaching are terms related to the reaction of dissolution of the metals. While (bio)leaching is desirable, (bio)corrosion is an undesirable phenomenon.

Liquid film flows coating a solid surface have received much attention in recent decades due to their vast industrial applications, such as surface protection, lubrication and cooling. The authors suggest several future research directions, including flow control and optimization and machine learning.

Additionally, theoretical and experimental studies on the generation of ultra-bright internal second harmonics are presented. A model based on one-dimensional nonlinear Maxwell curl equations without taking into consideration the slowly-varying envelope approximation has been developed.

In quantum mechanics and particle physics, Spin is considered as an intrinsic form of the quantum angular momentum of a point particle. As such, the authors aim to demonstrate that in accordance with the creative original idea of Kronig, Uhlenbek and Goudsmit, we can associate Spin with an intrinsic form of two angular momenta of the quantum Spherical Top.

The soft physical effects of various forms of shock waves are assessed in the context of molecular ensembles in liquids and polymers for the selective control of the energy state of its individual structural components.

In order to describe Hadron dynamics properly, the embedding of 4-dimensional space to 5-dimensional space is tried in lattice simulations, and in the light front holographic quantumchromo dynamics approach in which symmetric light-front dynamics without ghost are embedded in AdS5.

This compilation also examines gluons, vector gauge bosons that mediate strong interactions of quarks in quantum chromodynamics.

In closing, to solve nonlinear diffusion problems on a sphere, apart from the pole-bordering method, two implicit, balanced and unconditionally stable finite-difference schemes of the second and fourth approximation orders in spatial variables are proposed.

Chapter 1 - (Bio)corrosion and (bio)leaching are terms related to the reaction of dissolution of the metals. While (bio)leaching is desirable, (bio)corrosion is an absolutely undesirable phenomenon. Therefore, the researchers have been searching opportunities to increase bio(leaching) and to minimize, or avoid the process of (bio)corrosion. Application of electrochemical methods offers opportunity to determine the dissolution effect of chemical agents and microorganisms on metals, alloys, minerals and ores. The electrochemical methods can be used to quantitatively estimate the degree of corrosion metals and their alloys, or the degree of the dissolution of ores and minerals in the presence of various agents. The main goal of this chapter is to consider the explanation of (bio)corrosion and (bio)leaching on the base of changes of electric potentials.

Chapter 2 - Liquid film flows coating a solid surface have received much attention during last decades due to their vast industrial applications, such as surface protection, lubrication, cooling etc. To understand the dynamics of the film flows, involving instability, pattern formation, rupture, the long-wave models were used based on the fact that the film thickness is much smaller than a typical wave length. In literature studies, past researches focused on two cases: a liquid film flowing down a flat plane and a liquid film coating on a circular cylinder. In this review, we focus on the latter case: a liquid film on a circular cylinder. This flow is essentially unstable because of the famous Plateau-Rayleigh mechanism. Therefore, it is interesting to design active control methods to modulate the instability. Here, the influences of wall slippage, thermocapillary force and electric field will be reviewed. Finally, we will suggest several future research directions, including flow control and optimization and machine learning.

Chapter 3 - The chapter presents theoretical and experimental studies of the generation of ultra-bright internal second harmonic during femtosecond superradiant emission in multiple section semiconductor laser structures. Experimentally measured conversion efficiencies are by 1-2 orders of magnitude greater than those predicted by a standard SHG theory. To explain this fact, a model based on one-dimensional nonlinear Maxwell curl equations without taking into consideration the slowly-varying envelope approximation has been developed. The proposed model explains well all experimental data available to date. We show that the unique features of the superradiant emission in semiconductor media are responsible for the observed abnormal SHG. In particular, it has been demonstrated that strong transient periodic modulation of e-h density and refraction index dramatically affects the process of superradiance in semiconductor media. The presence of coherent non-equilibrium carrier density and refractive index gratings can explain the observed nonlinear effects, including the ultra-strong internal second harmonic generation and superluminal optical pulse propagation. A periodic grating rephases the nonlinear polarization and the generated electromagnetic waves. Strong periodic $\lambda/2$ modulation of the e-h density and corresponding modulation of the refractive index can result in quasi-phase matching conditions in a similar manner as in periodically poled materials for harmonics generation.

Chapter 4 - In quantum mechanics and particle physics, Spin is considered as an intrinsic form of the quantum angular momentum of a point particle. The subject of this chapter is to demonstrate that in accordance with the creative original idea of Kronig, Uhlenbek and Goudsmit, we can associate Spin with an intrinsic form of two angular momenta of the quantum Spherical Top. It is shown that the internal symmetry of the Intrinsic Top really exists and it manifests itself as an emergent property of the system of fundamental and simplest geometrical quantities. Thatiswhythisphenomenon is called Emergent Spin and to a particle with Emergent Spin one half we put into correspondence a spin field as a carrier space of internal symmetry of the Intrinsic Top. Three families: the electron and the electron neutrino, the muon and the muon neutrino, the tauon and the tauon neutrino can be considered as the evident experimental confirmation of the concept of Emergent Spin since from this new point of view they represent different states of a particle with Emergent Spin described by the spin field. The total number of these states equals four and, hence, one more state is predicted. The concept of Spin as an emergent property should be interesting in terms of discussion of possible ways to look for physics beyond the Standard Model since there is no doubt that new physics really exists and we need a clear guidance to the best place to look. To this end, we develop here new fundamentals for the particle physics on the ground of the concept of Emergent Spin and duality of natural Time predicted earlier.

Chapter 5 - The method of gravitational mass spectroscopy (GMS) was used to study the effect of weak shock waves (SW) on the formation of atomic nuclei concentration (ANC) in water, ethanoic acid, acrylamide, carbonic acid and protein (gelatin). The phenomenon of energy storage inside the ANC ensembles was discovered, and after the action of SW was switched off, a sharp release of energy in the form of a strong secondary SW "running through" all ANC in the ensemble and activation/destruction of the most unstable mass cluster. The selectivity of the process was achieved by selecting the intensity of the SW, as well as their shape and time of influence. The structures of some experimentally discovered clusters were presented.

Chapter 6 - Gluons are strong interaction gauge fields which interact between quarks, i.e., constituents of baryons and mesons. Interaction of matters is phenomenologically described by gauge theory of strong, electromagnetic, weak and gravitational interactions. In electro-weak theory, left handed leptons lL and neutrino νL, right handed leptons lR and left handed quarks uL,dL and right handed quarks uR,dR follow SU(2) ×

U(1) symmetry. Charge of leptons and quarks define hypercharge Y, and via Higgs mechanism SU(2)L ×U(1)Y symmetry forms U(1)em symmetry. Presence of JP = 3/2+ baryons, or N∗++ ∼ uuu suggests a new degree of freedom "color" for quarks, which follows SU(3) symmetry group. Hence the gluon fields are expressed as Aa μ where a = 1,2,··· ,8 specify the color SU(3) bases, and μ are 4-dimensional space-time coordinates. The quantum electrodynamics was extended to quantum chromo dynamics (QCD). Since ∂μAa μ is not a free field, the gauge theory requires ghosts that compensates unphysical degrees of freedom of gluons. Gluons, ghosts, leptons and quarks are related by Becchi-Rouet-Stora Tyuitin (BRST) transformation of electro-weak and strong interaction of the U(1)×SU(2)×SU(3) symmetric Faddeev-Popov Lagrangian. In order to describe Hadron dynamics properly, embedding of 4 dimensional space to 5-dimensional space was tried in lattice simulations, and in light front holographic QCD (LFHQCD) approach in which conformally symmetric light-front dynamics without ghost are embedded in dS5, and a parameter that fixes a mass scale was chosen from the Principle of Maximum Conformality. Coulomb or Landau gauge fixed Faddeev Popov Yang-Mills field equation is known to have the Gribov ambiguity, and tunneling between vacua between different topological structures was proposed by van Baal and collaborators. The symmetry of three colors can be assigned three vectors of quaternion H, whose multiplication on 2×2 matrices of Dirac spinors on S3 induces transformations. Instantons or sphalerons whose presence is expected from conformal equivalence of S3 ×R to R4 are reviewed. An extension of the dynamics embedded in complex projective space is proposed for understanding field theories.

Chapter 7 - Gluons are vector gauge bosons that mediate strong interactions of quarks in quantum chromodynamics (QCD). The gluon distribution is notoriously difficult to measure over the whole Bjorken-x range. In the small x region ($x \leq 0.1$), the gluon distribution is reasonably well measured by the HERA experiments. This does not help at larger x because the scaling violations due to quarks radiating gluons dominate over the gluon splitting contribution. A very strong constraint for the gluon distribution at medium and large x is, however, given by the energy momentum sum rule. At sufficiently small x, non-linear gluon interaction effects have been considered in order to moderate the rise of the cross section. The measurements of the structure functions by deep inelastic scattering (DIS) processes in the small x region have opened up a new era in parton density measurements inside hadrons. The structure functions reflect the momentum distribution of partons in a nucleon. Their steep increase towards small x, observed at HERA, also indicates a similar increase in gluon distribution towards small x in perturbative QCD. Therefore, gluon distribution is the observable that governs the physics of high-energy processes in QCD. Despite the difficulty to study their behavior, we shall try to have a closer look at gluons in this chapter.

Chapter 8 - Three methods for solving the problems of linear advection-diffusion and nonlinear diffusion on a sphere are proposed.

The first method is developed for both of these problems. The velocity field on the sphere is assumed to be non-divergent and known. Discretisation of the advection-diffusion equation in space is performed by the finite volume method using the Gauss theorem for every grid cell. For the discretisation in time the symmetrised dicyclic component-wise splitting method and the Crank-Nicolson schemes are used. The one dimensional periodic problems arising at splitting in the longitudinal direction are solved via Sherman-Morrison's formula and by Thomas' algorithm. The highlight of the method is the use of special bordered matrices for direct (i.e., non-iterative) solving the 1D problems arising at splitting in the latitudinal direction. The bordering procedure requires a prior determination of the solution at the poles. The resulting linear systems have tridiagonal matrices and are solved by Thomas' algorithm, which ensures the second approximation order in space and time. The method is thus implicit, unconditionally stable, non-iterative and computationally cheap. The theoretical results are confirmed numerically by simulating various linear advection-diffusion problems and nonlinear diffusion processes. The numerical tests show high accuracy and efficiency of the method that correctly describes the advection-diffusion processes and the mass balance of a substance in a forced and dissipative discrete system. In addition, in the absence of external forcing and dissipation, it conserves both the total mass and the norm (or energy) of solution.

To solve nonlinear diffusion problems on a sphere, apart from the pole-bordering method two implicit, balanced and unconditionally stable finite-difference schemes of the second and fourth approximation orders in spatial variables are proposed. Unlike the first, pole-bordering method, both nonlinear solvers exclude from consideration the polar grid cells with the poles, thereby avoiding the challenge of imposing suitable boundary conditions. The highlight of these two methods is the use of different coordinate maps of the sphere at each stage of splitting. As a result, all one-dimensional split problems are solved with periodic boundary conditions in both directions — in latitude and in longitude. In particular, all the 1D split problems in the first nonlinear method are solved using Sherman-Morrison's formula and Thomas' algorithm.

The common part of all the three methods is operator splitting. Due to the dicyclic coordinate splitting all the methods possess the second approximation order in time. The operator splitting also ensures direct and computationally cheap implementation of all the implicit schemes, as well as allows using parallel processors when solving the corresponding 1D split equations.

In: Horizons in World Physics. Volume 302 ISBN: 978-1-53617-180-8
Editor: Albert Reimer

Chapter 1

UNDERSTANDING ELECTRIC POTENTIAL IN PROCESSES OF (BIO)CORROSION AND (BIO)LEACHING

Biljana S. Maluckov*
Technical Faculty in Bor, University of Belgrade, Bor, Serbia

ABSTRACT

(Bio)corrosion and (bio)leaching are terms related to the reaction of dissolution of the metals. While (bio)leaching is desirable, (bio)corrosion is an absolutely undesirable phenomenon. Therefore, the researchers have been searching opportunities to increase bio(leaching) and to minimize, or avoid the process of (bio)corrosion. Application of electrochemical methods offers opportunity to determine the dissolution effect of chemical agents and microorganisms on metals, alloys, minerals and ores. The electrochemical methods can be used to quantitatively estimate the degree of corrosion metals and their alloys, or the degree of the dissolution of ores and minerals in the presence of various agents. The

* Corresponding Author's E-mail: nachevich@yahoo.com.

main goal of this chapter is to consider the explanation of (bio)corrosion and (bio)leaching on the base of changes of electric potentials.

INTRODUCTION

Metal corrosion represents a big problem in the industry. The solution to the problem of corrosion is focused on its reduction and corrosion inhibition. However, in some cases corrosion can occur despite the use of protective measures due to the presence of microorganisms which cause corrosion, which do not occurred in abiotic conditions (Maluckov 2012a, Maluckov 2012b). Corrosion which caused or promoted by microorganisms is called microbiologically influenced corrosion, microbiologically induced corrosion (MIC), microbiological corrosion or biocorrosion.

Corrosion is an electrochemical process which consists of the anode reaction that includes ionization (oxidation) of the metal (corrosion reactions), and the cathode reaction based on reduction of the chemical species (Beech and Gaylarde 1999). Metals initially oxidize to their ionic forms and release free electrons. The dissolved oxygen in the aqueous environment reacts with a water molecule and a free electron, forming hydroxyl ions. Hydroxyl anions then react with metal cations and form a corrosion product. In the absence of oxygen, the usual cathode reactions are the reduction of hydrogen ions or water (Paital and Dahotre 2009). A fundamental property of electrochemically based metallic corrosion is the equality of the rate of anodic, or oxidation reaction and the rate of the cathodic, or reduction reaction (Ratner et al. 2004). The degree of anode reaction (dissolution of the metal) decreases gradually with time, because the oxidation products (corrosion products) adhere to the surface forming a protective layer which provides a diffusion barrier for the reactants (Beech and Sunner 2004).

If microorganisms are present in aqueous solution, they may be reversibly associated with polymer-coated surface, or may initiate the process of irreversible adhesion by binding to the surface via exopolysaccharide glycocalyx polymer. The attached to surface microorganisms divide producing microcolonies within an adherent multispecies biofilm (Costerton et al. 1987). The bacterial adhesion forces to the metals are influenced by the electrostatic force and metal surface hydrophobicity (Sheng et al. 2007).

Corrosion and formed corrosion products are located from the surface to the solution, as a result of the gradual accumulation of inorganic passive layers on the surface of the metal. In contrast, the biofouling process of colonization and formation of biofilm by attached microorganisms on the surface and subsequently increase biofilm thickness via the growth of microorganisms and creation of the extracellular polymeric substances (EPS) are to the opposite direction (Videla 2001). The biofilm grows by internal replication and by recruitment from the bulk fluid phase (Costerton et al. 1987).

The oxidative dissolution of the mineral in nature under the influence of weather conditions can also be considered as an electrochemical reaction in which the oxidation agent is reduced and the sulfide ion is oxidized. Natural oxidation dissolution is used in hydrometallurgical process for the recovery of metals from ores, which is referred to leaching. They are added to various oxidants to improve the effect of leaching. When added microorganisms act as oxidants, process is named bioleaching.

In bioleaching, the microorganisms derive energy by oxidising the sulfur moiety and ferrous ion, which can be interpreted using electrochemistry and chemiosmotic theory (Hansford and Vargas 2001).

Good knowledge of electrochemical processes can enable improvement of (bio)leaching, and inhibition of (bio)corrosion in industry.

Analysis of changes in electric potential is a good method for monitoring and analysis of chemical and microbiological induced corrosion (Wang et al. 2006) of metals and alloys, as well as, for the analysis of (bio)leaching of ores and minerals (Niu et al. 2014) in nature and industry.

For investigation of electrochemical corrosion are usually applied the open circuit potential (OCP), cyclic voltammetry (CV), potentiodynamic polarization (PP), chronoamperometry (CA) and electrochemical impedance spectroscopy (EIS).

In addition, electrochemical noise analysis (ENA), electrochemical quartz crystal microbalance (EQCM), microelectrode techniques (Wang et al. 2006) and scanning vibrating electrode technique (SVET) (Iken et al. 2008) can be used for studying of the microbially influenced corrosion process.

For a quick estimation of the influence of an agent on the dissolution of metal, alloy and mineral can use the potentiodynamic polarization, and for the evaluation of the mechanism of corrosion should be used the combinations of various electrochemical and analytical methods.

Microscopically obtained data on the type of microorganisms, combined with the data from the chemical analysis of surfaces, and electrochemical measurements provide information about the chemical composition of the corrosion products and microbiological deposits, can be used to estimate the level of corrosion (Maluckov 2012b). In the chapter considers how can be explained processes of (bio)corrosion and (bio)leaching, in general and on some concrete practical examples, on the basis of electric potential changes.

1. Understanding Electric Potential of (Bio)Corrosion

Metal corrosion in the industry leads to long-term deadlock of industrial processes and high costs of reparation of damaged plants. Therefore, the investigations are directed towards finding solutions how to reduce or eliminate corrosion processes in the industrial installations. This is particularly interesting regarding corrosion effects on steel which is the most common used material for construction the industrial plants. In addition, the copper is interesting because it has application in various industries, and aluminium is the most common used material in the aero industry.

Investigation of titanium-based material for medical purposes is actual due to their biocompatibility and resistance on corrosion (Maluckov 2014).

EPS and other excretion metabolites of microorganisms can affect the electrochemical properties of the metal surface and play an important role in corrosion of metals.

EPS are mainly composed of polysaccharides and proteins (Fang et al. 2002, Kinzler et al. 2003). Metal binding of EPS involves the interaction between anionic functional groups (eg carboxyl, phosphate, sulfate, glycerate, pyruvate and succinate groups) and metal ions. The affinity of anionic ligands for multivalent ions, such as Ca^{2+}, Cu^{2+}, Mg^{2+}, and Fe^{3+}, can be strong and can lead to a significant decrease in standard potential. For example, for Fe^{3+}/Fe^{2+} the redox potential varies significantly with respect to different ligands (from + 1.2 V to -0.4 V). EPS binding metal ions, open new redox reactions in the biofilm/metal system, such as the direct transfer of electrons from metals (e.g., iron) or biominerals (e.g., FeS). In the presence of suitable electron acceptor (e.g., oxygen in the anaerobic system, or nitrates under anaerobic conditions), depolarization of the cathode and acceleration of corrosion of the mechanism (Beech and Sunner 2004) is caused.

1.1. Electrochemical Studies of Corrosion Processes with Microorganisms Isolated from Industrial Plants

Acording to Starosvetsky et al. 2007 uncovering of MIC in technological equipment failures requires an individual approach for each case and that in certain cases, assessment role of microorganisms in corrosion destruction is possible only by simulation of the corrosion parameters found in the field, which have a real effect on the process.

Corrosion process induced by sulfate-reducing bacterium, *Desulfovibrio alaskensis* (strain IMP-7760), isolated from a water oil separator at the Samaria Pol II complex (Gulf of Mexico) on SAE1010 carbon steel was examined by analyzing electrochemical noise under nutritionally rich and oligotrophic growth conditions. Studies have shown that in nutritively rich conditions, steel was first exposed to uniform corrosion, then in a short period to localized corrosion with values of the localization index LI close to 0.3, and finally to the mixed corrosion processes. In contrast, under oligotrophic conditions, there were fluctuations among the three mechanisms of corrosion during the first 100h, then a long period of localized corrosion with LI values reaching 1, and finally the corrosion process has become uniform (Padilla-Viveros et al. 2006).

The corrosion process was investigated for two steel alloys St-35.8 (a typical carbon steel alloy employed in European naval industry) and API-5XL52 (a weathering steel alloy employed in Mexico oil facilities) induced by a hydrogenotrophic sulfate-reducing bacterium (SRB) *Desulfovibrio capillatus* (DSM14982T). *Desulfovibrio capillatus* (DSM14982T) was isolated from a water/oil mixture from Samaria III oil field separator with serious corrosion problems, near the Gulf of Mexico. The polarization resistance R_p has reached values which are lower in samples with inoculum than in the control sterile medium. The change in the aggressiveness of the solutions on the metal surfaces was reflected in the decrease of E_{corr} with an accompanying decrease of R_P. Decreasing of E_{corr} and R_P indicate that bacteria could improve the anode reaction (Miranda et al. 2006).

Starosvetsky et al. 2008 have made an analysis of localized electrochemical corrosion mechanism of stainless steel (SS) types UNS S30403 and UNS 31603 in the presence of iron-oxidizing bacteria *Sphaerotilus* sp. isolated from the rust deposits of carbon steel of the clogged heat exchanger from the oil refinery plant (Haifa, Izrael). There was no evidence of pitting corrosion of SS samples after exposure to sterile 3% NaCl, while in the presence of iron oxidizing bacteria, numerous pits were detected on samples of 304 L steel exposed to iron hydroxide sedimentary layers. Steel 316L showed more resistance to pitting corrosion than steel 304L. The results revealed an instantaneous gradual shift of the transient potential of both steels towards negative potentials from steady-state value of -0.15 V to -0.35 to -0.42 V (SCE) during the whole exposure interval since IOB culture addition into sterile 3% NaCl solution (Starosvetsky et al. 2008).

Xu et al. 2008 have been investigating pitting corrosion behavior of stainless steel 316L SS in the presence of aerobic iron-oxidizing (*Leptothrix* sp.) and anaerobic sulfate-reducing bacteria (*Desulfovibrio* sp.) isolated from cyclic cooling water system of an oil refinery plant. The results show that the corrosion potential (E_{corr}), pitting potential (E_{pit}) and polarization resistance (R_P) of 316L SS had a distinct decrease in the presence of bacteria, in comparison with those observed in the sterile medium for the same exposure time interval. The addition of chloride ions further accelerates the decline range in the values of E_{corr}, E_{pit} and R_P, and induces localized corrosion (Xu et al. 2008).

Studies of the corrosion of austenitic steel type SS304L have shown that the presence of uniform corrosion in contact with sugar cane juice is low, and that it depends on the pH of the juices. Pitting corrosion mainly occusr in fresh or old juice, but not in the sterilized juice. Corrosion is increased under the passive film and there were small craters on the surface of the samples. Results of changes of the corrosion potential (E_{corr}) have shown that when the pH value of the juice is reduced, the sample SS304L becomes less resistant to corrosion. This is suggested that should be shorten of the contact time with the given juice to prevent the effects of their aging (Durmoo et al. 2008).

The study of the effects of tequila on corrosion process of aluminum, copper, brass and Stainless Steel 304 (SS 304) by potentiodynamic polarization and Electrochemical Impedance Spectroscopy (EIS) measurements showed that the corrosion rate values follow the trend Al > Cu > brass > SS 304 (Carreon-Alvarez et al. 2012).

The influence of hydrocarbon-degrading bacterial biofilm *Bacillus cereus* ACE4 and *Serratia marcescens* ACE2 strains on aluminium AA 2024 aeronautical alloy exposed to fuel/water mixtures in 1% NaCl solution was examined by EIS. Bacterial strains were isolated from corrosion products in diesel-transporting pipelines in a northwestern region of India. Studies EIS show that *B. cereus* ACE4 causes more severe pitting corrosion in AA 2024 alloy than *S. marcescens* ACE2 in the fuel/water system (Rajasekar and Ting 2010).

1.2. Electrochemical Studies of the Environmental Impact on Biocorrosive Processes

The electrochemical behavior of stainless steels (SS) in natural waters is characterized by the ennoblement of their free corrosion potential (E_{corr}) (Landoulsi et al. 2008a). In synthetic fresh water (SFW) the ennoblement of the free corrosion potential (E_{corr}) of AISI 31stainless steel (Landoulsi et al. 2008b) did not occur. According to Videla 1994 biofilms can induce localized corrosion on stainless steel in seawater by creating differential aeration effects and induction of drastic changes in the local chemistry of the metal/biofilm/solution interfaces which ennoblement of the corrosion potential.

The SRB in the biofilm was aggregated into clusters, and caused increasing of the production of EPS when was exposed to seawater media containing toxic metals and chemicals. The microbial clusters became barriers to diffusion and were located in the area under the cathode. The regions between microbial clusters allow the surface to reach greater access to chloride and sulfate in the seawater medium, and to act as the anode, resulting in accelerating the electrochemical corrosion reactions (Fang et al. 2002).

The colonization of a marine aerobic *Pseudomonas* bacteria on the coupon surface of 304 stainless steel induced subtle changes in the alloy elemental composition in the outermost layer of surface films in a seawater-based medium. The most significant feature resulting from microbial colonization on the coupon surface was the depletion of iron and the enrichment of chromium content as compared to a control coupon exposed to the sterile medium, and the enrichment of chromium increased with time (Yuan and Pehkonen 2007).

Antony et al. 2007 investigated corrosion of 2205 duplex stainless steel in chloride medium containing sulfate-reducing bacteria *Desulfovibrio desulfuricans*. The E_{corr} drop was an indication of reducing nature of the environment due to the production of H_2S by SRB activity. The occurrence of the rapid growth of the current with formation of an active current peak, and the moderate-shifted potentials of E_{corr}, can be attributed to the formation of metal sulfides by reductive dissolution of oxides present in the passive film. It was also found that sulfide passive film depolarizes cathode reactions. Hence, the decrease in the current by further increase of the potential during the anode polarization are consequences of the decrease of the net current (anode-cathode) and not the anode current (Antony et al. 2007).

Thermal aging of 2205-Type DSS (UNS S31803) duplex stainless steel (DSS) can create more deleterious changes in the microstructure and hence undergoes greater extent of bacteria attack in 3.5% NaCl-based medium (Antony et al. 2010).

Electrochemical investigation of the effect of *Desulfovibrio desulfuricans* bacteria the DSM642 standard soil strain on passivity on 2205 duplex stainless steel in aerated 3.5% NaCl showed that the DSS corrosion resistance was affected in the presence of DSM642 biofilm. This statement was based on the significant decrease in the charge transfer resistance (R_1) obtained from EIS and the increase in the measured current densities obtained from potentiodynamic polarization curves (Dec et al. 2016).

Zhao et al. 2007 investigated the corrosion behavior of mild steel in the sea mud with active SRB. The value of R_p increases at the first stage, decreases in the second stage with the variation of time, and increases again at the third stage. The metabolic activity of SRB and the change of state of the surface on the steel can be the explanation dependence of polarization resistance R_p on time. In the environment of sea mud containing SRB, the original corrosion products, ferric (oxyhydr) oxide, transformed to FeS which was decreased the corrosion rate to some extent. With the proliferation of SRB and excess of the dissolved H_2S, the composition of the protective layer formed of FeS transformed to FeS_2 or other non-stoichiometric polysulfide, which changed the state of the former layer and accelerated the corrosion process. At the third stage, the metabolic activity of the microorganisms declined gradually, protective layer of corrosion products is formed again and the corrosion rate decreases (Zhao et al. 2007).

In order to obtain more information on impact of environmental on the bio(corrosion) of mild steel, studies were carried in lactate/sulfate and lactate/nitrate media. The values of redox potential in Table 1 show changes in the negative direction due to SRB activity in both media. The differences between values of E_{redox} in both sterile media can be explained that the lactate/sulfate medium already contaminated with sulfide which provide a very reducing environment, whereas in the lactate/nitrate medium the bacterial growth is ensured only by deaeration. Corrosion current density shows an enhancement due to the SRB in the lactate/sulfate medium but not in the lactate/nitrate medium (Fonseca et al. 1998).

Table 1. Redox potentials for different investigated media

Medium	E_{redox} (mV vs SCE)
Lactate/nitrate	-167
SRB in lactate/nitrate	-479
Lactate/sulfate	-362
SRB in lactate/sulfate	-444

Fonseca et al. 1998

The development of biofilms in water copper pipes can cause not only microbiological malfunction of water, but also to the appearance of corrosion, which is different from the usual types copper corrosion (Maluckov 2013).

Yuan et al. 2007 investigated the corrosion behavior of 70/30 Cu-Ni alloy in the absence and presence of marine aerobic *Pseudomonas* NCIMB 2021 bacteria in nutrient-rich simulated seawater-based medium as a function of exposure time. Polarization curve measurements showed that the corrosion current density $i_{corrosion}$ of the alloy specimens in the presence of *Pseudomonas* was higher by as much as 14-fold as compared to those of the control specimens after 42 days. Corrosion potentials, $E_{corrosion}$, underwent a slightly positive shift after 42 days of immersion in the sterile nutrient-rich medium, may be caused by the inhibition effect of the protective oxide film on the electrode reactions. In the *Pseudomonas* inoculated medium occur negative shift in the corrosion potential probably indicates the acceleration of corrosion. The EIS measurements showed that the charge transfer resistance, R_{ct}, and the resistance of oxide film R_f, gradually increased with time in the abiotic medium and dramatically decreased with time in the biotic medium inoculated with the *Pseudomonas,* which indicate of the acceleration of corrosion rates of the alloy in presence bacteria (Yuan et al. 2007).

Liu et al. 2018a compared two types of antibacterial Cu-bearing 316L stainless steel (with the addition of rare earth elements 316L-Cu-A and without rare earth elements 316L-Cu-B) with stainless steels in the presence of sulfate-reducing bacteria. The passive currents density in potentiodynamic polarization measurements indicate that the two types of Cu-bearing 316L SS have a better anticorrosion performance compared to the control 316L SS in the absence of SRB in the medium, and 316L-Cu-A is better than that of 316L-Cu-B. The OCP was the most negative for the control 316L SS. The Cu bearing steels have less negative OCP, but the OCP of 316L-Cu-A is more negative than that of 316L-Cu-B in the SRB culture medium. The smallest average R_p for the 316L-Cu-B indicate the highest corrosion rate in the SRB culture medium. From EIS and potentiodynamic polarization measurements, more serious pitting corrosion and uniform corrosion occur on 316L-Cu-A and 316L-Cu-B compared with 316L SS in the SRB environment. The uniform and pitting corrosion of 316L-Cu-B are worse than those of 316L-Cu-A (Liu et al. 2018a).

The results of the investigation of adhesion of two anaerobic sulfate-reducing bacteria (*Desulfovibrio desulfuricans* and a local marine isolate) and an aerobe (*Pseudomonas* sp.) bacteria to four polished metal surfaces (i.e., stainless steel 316, mild steel, aluminium, and copper) showed that the bacterial adhesion force is the highest on aluminium, and smallest on copper. The adhesion to the metals by *Pseudomonas* sp. and *D. desulfuricans* was greater than by the marine isolated SRB (Sheng et al. 2007).

1.3. Enzymatic Electrochemical Tests of Biocorrosion

The release of extracellular enzymes (polysaccharidases, proteases, lipases, esterases, peptidases, glycosidases, phosphatases and oxidoreductases) by microorganisms into their external environment provides the basis for the interaction between cells and the substrate (Beech et al.2005).

All the identified enzymes responsible for microbial influenced corrosion of stainless steels belong to the class of oxidoreductases. Their ability to exchange electrons seems to be crucial for their involvement in microbial influenced corrosion. The increase of the cathode reaction rate induced by enzymes could occur only in a restricted range of redox potential inside biofilms which may explain why microbial influenced corrosion does not always occur when bacteria are present on metallic materials (Landoulsi et al. 2008a).

Iken et al. 2008 used a scanning vibrating electrode technique (SVET) to investigate the catalysis of oxygen reduction in the presence of hema protoporphyrin (hemin) enzymes in the aerobic medium. Based on the SVET analysis, it has been shown that hemin provokes the appearance of a galvanic cell on the surface of the material leading to localized corrosion. The results showed that the hemin adsorbed on the surface of a sample of stainless steel led to an increase in cathode current, indicating a catalysis of oxygen reduction. On the SVET mape, zone without hemin differed from hemin modified zone in at least four times the negative current (Iken et al. 2008).

Landoulsi et al. 2008b are used glucose oxidase and glucose (G_{ox}) for electrochemical tests of biocorrosion AISI316L stainless steel in synthetic fresh water (SFW). In G_{ox} media, the E_{corr} ennoblement occurs and reaches values much higher than in SFW and the concentration of H_2O_2 increased progressively. The E_{corr} ennoblement is probably directly due to the presence of H_2O_2 and to the electrochemical behavior of H_2O_2 or related oxygen species. In G_{ox} medium, the surface oxide layer is not further enriched in Fe^{ox} compared to Cr^{ox} and the disappearance of the colloidal particles, due to dissolution of ferric hydroxide by complex formation with gluconate (Landoulsi et al. 2008b).

2. Understanding Electric Potential of (Bio)Leaching

The anodic dissolution of the mineral as a semiconductor can generally be described by the following reactions where X and Y represent ionic species in solution (Crundwel 1988):

$$(M^{\delta}+ S^{\delta}-)_{s} + X \rightarrow (MX)+ + \bullet S_s + e- \quad (1)$$

$$(M^{\delta}+ S^{\delta}-)_{s} + X + h+ \rightarrow (MX)+ + \cdot S_s \quad (2)$$

$$\bullet S_s + Y \rightarrow (SY)^{+} + e- \quad (3)$$

$$\bullet S_s + Y + h + \rightarrow (SY)^{+} \quad (4)$$

$$\bullet S_s + \bullet S_s \rightarrow (S\text{-}S)_{s}. \quad (5)$$

$$S_n + \bullet S_s \rightarrow S_{n+1} \quad (6)$$

The life of microorganisms occurs roughly between -500 mV and +800 mV (Sand 2003). The bioleaching conditions typically exhibit a relatively high redox potential around 0.65-0.70 V (SHE) (Watling 2006).

In study of bioleaching using iron and sulfur oxidising microorganisms and heterotrophs enriched from self-heating pyritic coal, rapidly increased ORP and reduction in pH subsequent to inoculation point to the development of bacterial communities active in the key bioleaching functions of the oxidation of Fe^{2+} to Fe^{3+} and the oxidation of reduced inorganic sulfur species (Watling et al. 2013). Once initiated, pyrite oxidation was self-sustaining because Fe^{3+} oxidized the sulfide of the pyrite being reduced to Fe^{2+}, and due oxygen, in the presence of bacteria, oxidized Fe^{2+} to Fe^{3+} at the pyrite/solution interface (Cabral and Ignatiadis 2001). The behavior of attached bacteria is very dependent on the Fe^{+3}/Fe^{+2} ratio in the exopolymer membrane, which itself depends on the redox potential in solution and the concentration of dissolved iron (Hansford and Vargas 2001). It is reported in the literature that the standard potential $E^0_{Fe^{3+}/Fe^{2+}}$ in a sulfate and acid medium is 680 mV(NHE), or about 440 mV(SCE) (Cabral and Ignatiadis 2001). When the redox potential E_{EP} (the redox potential of the complexed Fe^{+3}/Fe^{+2} in the EPS layer of attached bacteria, which should evolve closely to the solution redox potential) is minimized, which is obtained when most of the dissolved iron in solution is present as ferrous ion, potential difference between E_{EP} and E_{ox} (the redox potential of the O_2/H_2O couple in the cytoplasmic space, which approaches + 0.82 V (SHE) in oxygenated solution) is maximized and most of the bacterial activity is linked to ferrous ion oxidation (Hansford and Vargas 2001). Because of the block of cell envelope and the low difference of redox potential between the intracellular and extracellular surroundings, the proceeding of extracellular electron transfer (EET) depends mainly on the help of a variety of mediators that function as an electron carrier or bridge (Liu et al. 2018b).

In the EPS film layer can concentrate Fe^{3+} and H^+ ions. The EPS film layer with Fe^{3+} deposits on the surface of chalcopyrite becomes a barrier of oxygen transfer to chalcopyrite and passivate chalcopyrite, at the same time, can creates the high redox potential space through concentrating Fe^{3+} ions, especially more than 750 mV (SHE), to accelerate bioleaching pyrite in chalcopyrite concentrates (Run-lan et al.2008).

According to Cabral and Ignatiadis 2001, in bioleaching, the overall process of pyrite dissolution displays a rapid acceleration phase, followed by deceleration mainly due to the deceleration and subsequent termination of the oxidation of $Fe^{2+}$to Fe^{3+} by oxygen in the presence of the microorganisms. Pyrite oxidation began when the potential acquired by the pyrite fell below that of the platinum. The separation point was close to the redox potential of the Fe^{3+}/Fe^{2+} couple (≂0.68 V (NHE) at 25^{0}C in H_2SO_4 medium). The Fe^{2+} disappeared completely from the solution. The difference in potential (overpotential between $E_{platinum}$ and E_{pyrite}) amounts to a negative polarization of the pyrite, which places it in a reducing pyrite position with respect to the strong oxidant, Fe^{3+}, present in solution. This difference in potential increased due to the continuous rise in the Fe^{3+} content in solution. An excessive Fe^{3+} concentration imposing an excessive redox potential on the solution and thereby restricting bacterial activity (above a redox potential of 840 mV(NHE)), despite pyrite oxidation by Fe^{3+}. This permits the protection of the pyrite by Fe^{2+}, and the growing inaccessibility of the pyrite surface covered by jarosite scale occur termination of the oxidation of Fe^{2+} to Fe^{3+} by oxygen in the presence of the microorganisms (Cabral and Ignatiadis 2001)

Electrochemical investigations showed that oxidation of pyrite occurred by two reaction pathways. The first pathway occurred at low potential (0.50V) electrochemical oxidation of pyrite was diffusion-limited due to a sulfur-rich layer (S^0) with of Fe $(OH)_3$ that formed and covered the pyrite surface, resulting in surface passivation. As the potential increased to 0.60 V, the primary sulfur species, amorphous S_8, was converted to crystalline S_8, which reduced the surface passivation and allowed for the continued oxidation of pyrite. At potentials of 0.70 and 0.80 V, the second reaction pathway of pyrite oxidation resulting in the generation of more $Fe(OH)_3$, Fe_2O_3, FeO and S_8 (Tu et al. 2017).

When arsenic is presents in the sulfide ore in the absence of bacteria, the concentration of Fe^{3+} ions decrease and at lower E_h potentials (<450 mV vs Ag/AgCl), As^{5+} is reduced cathodically, while dissolution of pyrite is the only remaining anode reaction. In aerobic conditions and in the presence of microorganisms, Fe^{2+} ions are easily reoxidized and the conditions of high potential are maintained (Wiertz et al. 2006). The electrochemical results and XRD analysis anodic process of arsenopyrite with bacteria showed that arsenopyrite is firstly oxidized to As_2S_2 at the potential of 0.2-0.3 V (SHE) and As_2S_2 covers the electrode and retards the process continuously. While at higher potential over 0.3 V (SHE), As_2S_2 is oxidized to H_3AsO_3, and H_3AsO_3 is then oxidized to H_3AsO_4 at 0.8 V (SHE) (Tao et al. 2008):

$$FeAsS \rightarrow Fe^{2+} + (1/2)As_2S_2 + 2e^- \quad (7)$$

$$As_2S_2 + 14H_2O \rightarrow 2H_3AsO_3 + 2SO_4^{2-} + 22H^+ + 18e^- \quad (8)$$

$$H_3AsO_3 + H_2O \rightarrow H_3AsO_4 + 2H^+ + 2e^- \quad (9)$$

2.1. The Effect of Change of the Electric Potential with Change Fe^{3+}/Fe^{2+} Ratio and Temperature

Comparing the thermophilic and mesophilic bio(leaching) conditions can spot differences in the oxidation–reduction potential. The increase in potential is usually the result of an increase Fe^{3+}/Fe^{2+} by activity microorganisms.

At thermophilic conditions the values *E*(SCE) are lower in comparison to mesophilic conditions. The lower redox conditions for the thermophilic bioleaching mean higher concentration of soluble ferrous ion in the leach liquor compared to the mesophilic bacterial leach condition column. The thermophilic column yielded higher ferrous ion concentration with a Fe^{3+}/Fe^{2+} molar ratio of 0.7 in the final pregnant solution, while the ambient mesophilic column yielded lower ferrous ion concentration with a Fe^{3+}/Fe^{2+} molar ratio of 100+ (Acar et al. 2005).

During biooxidation of refractory ore for 5 days at 42°C, there was a progressive increase in the Fe^{3+}/Fe^{2+} ratio and redox potential to a maximum value above 600 mV over 5 days, which indicates that ferrous iron present in the sulfide minerals, was oxidized to the ferric state (Amankwah et al. 2005).

Electrochemical tests have shown that it occurs at low temperatures 25°C of the anodic polarization curve the "passive" region situated within the potential range 0.45–0.65 V (SCE) where the current did not increase with increasing potential, while at 65°C there was no such evidence. Cathodic characteristics showed that the passive layers formed at 25°C and high potentials strongly inhibit ferric reduction on polarized $CuFeS_2$ surfaces (Tshilombo et al. 2002).

Metal-deficiency sulfides chalcocite and bornite were first formed with a low redox potential value (360–461 mV (SCE)), and then gradually transformed to covellite with a high redox potential value (461–531 mV(SCE)) during bioleaching of chalcopyrite at thermophilic conditions (Liu et al. 2016).

The addition activated carbon could significantly promote the dissolution of chalcopyrite for both bioleaching and chemical leaching at 65°C. Activated carbon could change transition path of electrons through galvanic interaction at a low redox potential (<400 mV(SCE)) to form more readily dissolved secondary mineral chalcocite and finally enhanced the copper dissolution (Ma et al. 2017).

2.2. The Effect of Applied Electric Potential and Temperature on Bio(Leaching)

At 75°C and E_h of 650 mV (SHE), the presence of pyrite compared to the same amount of the inert mineral of quartz, increases chalcopyrite dissolution rate by more than five times due to the effective galvanic interaction between chalcopyrite and pyrite. In addition, chalcopyrite leaching of Cu and Fe was found to be stoichiometric suggesting insignificant pyrite dissolution. Although the chalcopyrite dissolution rate at 750 mV (SHE) was approximately four-fold greater than at 650 mV (SHE) in the presence of pyrite, the galvanic interaction between chalcopyrite and pyrite was negligible. Approximately all of the sulphur from the leached chalcopyrite was converted to S^0 at 750 mV, regardless of the presence of pyrite. At this E_h approximately 60% of the sulfur associated with pyrite dissolution was oxidised to S^0 and the remaining 40% was released in soluble forms, e.g., SO_4^{2-} (Li et al. 2015).

Studies have shown that when controlled at redox potential 750 mV(SHE), and temperature changes from 35-75° C that the addition of aqueous iron plays an important role in accelerating Cu leaching rates, especially at lower temperature, primarily by reducing the length of time of the initial surface chemical reaction controlled stage (Li et al. 2016).

For the low-grade of chalcopyrite concentrate investigations have shown that without redox control, using a thermophilic culture at 70°C copper recoveries 95%. If moderately thermophilic microorganisms are used at 45°C, under controlled redox conditions, a leaching favors at the redox potential of about 420 mV copper recoveries can be increased from 64% at high redox potential to 97% and the pyrite was not oxidized. A key advantage of the process at 45°C is that the bioleaching process costs are minimized, since the pyrite was not oxidized (Gericke et al. 2010).

Dissolution of copper from high grade chalcopyrite flotation concentrates is the most efficient by the electrochemical bioleaching compared to the chemical leaching, electrochemical leaching and bioleaching. The electrochemical bioleaching with the control of redox potential between 400 and 425 mV (Ag/AgCl) increased up to 35% more copper recovery compared to the conventional bioleaching (Ahmadi et al. 2010). Under this condition, the cell concentrations increase drastically and significantly is reduced the formation of jarosite wich is known as a barrier for chalcopyrite dissolution. This leads to a higher electrochemical reduction of chalcopyrite and its improved dissolution (Ahmadi et al. 2010, Ahmadi et al. 2011). The addition of pyrite to the chalcopyrite concentrate significantly increased the efficiency of copper extraction in the electro-biochemical system. At low levels of solution ORP, pyrite remains inert, which acts as a cathode site relative to chalcopyrite and other copper sulfide minerals (galvanic interaction) leading to enhance the anodic dissolution of the copper bearing minerals. Electrochemical system regulates the ratio of Fe^{3+}/Fe^{2+} at an optimum level where the dissolution rate of chalcopyrite is maximum. Sulfur oxidizer microorganisms intensify the galvanic interactions and the rate of electron transfer among sulfides by removing the insulating sulfur product (Ahmadi et al. 2012).

By choosing the appropriate microbial community (Christel et al. 2018) it is possible to favor redox potential for the bioleaching of chalcopyrite.

3. Bio(Corosion) and Bio(Leaching) in Function of Obtaining New Materials

Bio(corrosion) and bio(leaching) are of great interest for engineering novel materials. Simultaneously with the processes of bio(corrosion) and bio(leaching), biomineralization occurs, during which new desirable materials may arise.

Microbial polysaccharides can template assembly of nanocrystal fiber (Chan et al.2004). Magnin et al. 1994 exposed the sintered iron-titanium hot-pressed plates to bacterial corrosion in the acidic environment to produce a porous material to infiltrate the lithium and used as an anode for thermal batteries. The corrosion potentials were very different in the bacteria-free and the inoculated electrolyte. In the absence of bacteria, the potential remained in the same range as that of a pure iron electrode, i.e., in the hydrogen evolution domain, with a slightly lower value presumably due to the catalytic effect of hydrogen evolution on titanium. The pH increase due to the cathode reaction caused the observed $Fe(OH)_2$ precipitation which lower the corrosion rate. Conversely, in the presence of bacteria, the corrosion potential rapidly drifted to more anodic values, precluding hydrogen evolution. Increasing the corrosion rate due to bacteria can be explained by the rapid bacterial oxidation Fe^{2+} produced by anodic reaction (Magnin et al. 1994). The interior of the copper tube pitting caused by microbial induced corrosion was covered by a green layer of malachite (Burleigh et al. 2014), and malachite is used in making jewelry. Coupling biological oxidation of Fe^{2+} by *Acidithiobacillus ferrooxidans* and chemical reduction Fe^{3+} by zero valent iron (ZVI) promotes biomineralization in simulated acidic mine drainage (Wang et al.2019). In such a system, schwertmannite was formed with little co-precipitation of other metals (Wang et al.2019) which could be used for environmental remediation (Zhang et al. 2018). Covellite (CuS) particles were synthetized from an AMD sample after 48 and 96 h of exposure to hydrogen sulfide (H_2S) produced in a bioreactor containing acidophilic sulfate reducing bacteria (SRB) (Silva et el. 2019).

CONCLUSION

By controlling the potential, it is possible to inhibit or favor certain electrochemical processes on metals, alloys and minerals.

In abiotic electrochemical corrosion, it is possible by control the potential in the area where the metal does not corrode to prevent the corrosion of metals and alloys, while by controlling temperature and potential, the passivation of sulfide minerals can be reduced.

If the potentials are maintained outside of the extent survival of microorganisms, the inhibition of metal corrosion induction by microorganism occurs, i.e., dissolution comes down to only on electrochemical processes, while maintainance of optimal conditions for growth and reproduction of microorganisms can increase dissolution of minerals and the yield of desired metals:

inhibition of biocorosion
/
$\Delta E_{forlife\ moo} > E > \Delta E_{forlife\ moo}$
\
inhibition of bioleaching

The thermophilic bioleaching potential is lower than the mesophilic bioleaching one. Because of all these mentioned above, electrobioleaching shows better properties than just bioleaching, because the control of the potential provides maximum conditions for growth and evolvence of microorganisms.

By choosing the appropriate microbial community it is possible to favor redox potential for the bioleaching, too.

In addition to everything, by maintaining certain potentials can be enabled bio(mineralization) and obtained desired materials.

REFERENCES

Acar, S., Brierley, J.A. and Wan, R.Y. (2005). Conditions for bioleaching a covellite-bearing. *Hydrometallurgy*, 77: 239–246.

Ahmadi, A., Schaffie, M., Manafi, Z. and Ranjbar, M. (2010). Electrochemical bioleaching of high grade chalcopyrite flotation concentrates in a stirred bioreactor. *Hydrometallurgy*, 104: 99–105.

Ahmadi, A., Schaffie, M., Petersen, J., Schippers, A. and Ranjbar, M. (2011). Conventional and electrochemical bioleaching of chalcopyrite concentrates by moderately thermophilic bacteria at high pulp density. *Hydrometallurgy*, 106: 84 -92.

Ahmadi, A., Ranjbar, M. and Schaffie, M. (2012). Catalytic effect of pyrite on the leaching of chalcopyrite concentrates in chemical, biological and electrobiochemical systems. *Minerals Engineering*, 34:11-18.

Amankwah, R.K., Yen, W.-T. and Ramsay, J.A. (2005). A two-stage bacterial pretreatment process for double refractory gold ores. *Minerals Engineering*, 18: 103-108.

Antony, P.J., Chongdar, S., Kumar, P. and Raman, R. (2007). Corrosion of 2205 duplex stainless steel in chloride medium containing sulfate-reducing bacteria. *Electrochimica Acta*, 52: 3985–3994.

Antony, P.J., Singh Raman, R.K., Raman, R. and Kumar P. (2010). Role of microstructure on corrosion of duplex stainless steel in presence of bacterial activity. *Corrosion Science,* 52: 1404–1412.

Beech, I.B. and Gaylarde, C.C. (1999). Recent advances in the study of biocorrosion - an overview, *Revista de Microbiologia*, 30: 177-190.

Beech, I.B. and Sunner J. (2004). Biocorrosion: towards understanding interactions between biofilms and metals. *Current Opinion in Biotechnology*, 15: 181–186.

Beech, I.B., Sunner, J.A. and Hiraoka, K. (2005). Microbe-surface interactions in biofouling and biocorrosion processes. *International Microbiology*, 8: 157-168.

Burleigh, T.D., Gierke, C.G., Fredj, N. and Boston, P.J. (2014). Copper Tube Pitting in Santa Fe Municipal Water Caused by Microbial Induced Corrosion. *Materials*, *7*(6): 4321-4334.

Cabral, T. and Ignatiadis, I. (2001). Mechanistic study of the pyrite–solution interface during the oxidative bacterial dissolution of pyrite (FeS_2) by using electrochemical techniques. *International Journal Mineral Processing*, 62: 41–64.

Carreon-Alvarez, A., Valderrama, R.C., Martínez, J.A., Estrada-Vargas, A., Gómez-Salazar, S., Barcena-Soto, M., Casillas, N. (2012). Corrosion of Aluminum, Copper, Brass and Stainless Steel 304 in Tequila. *International Journal of Electorchemical science*, 7: 7877 – 7887.

Chan, C.S., De Stasio, G., Welch, S.A., Girasole, M., Frazer, B.H., Nesterova, M.V., Fakra, S. and Banfield, J.F. (2004). Microbial polysaccharides can template assembly of nanocrystal fiber. *Science*, 303:1656-1658.

Costerton, J.W., Cheng, K. -J., Geesey, G.G., Ladd, T.I., Nickel, J.C., Dasgupta, M. and Marrie, T.J. (1987). Bacterial biofilms in nature and disease. *Annual Reviews of Microbiology*, 41: 435-64.

Christel, S., Herold, M., Bellenberg, S., Buetti-Dinh, A., El Hajjami, M., Pivkin, I.V., Sand, W., Wilmes, P. Poetsch, A., Vera M., and Dopson. M., (2018). Weak Iron Oxidation by *Sulfobacillus thermosulfido-oxidans* Maintains a Favorable Redox Potential for Chalcopyrite Bioleaching. *Frontiers in Microbiology*, 9:1-12.

Crundel, F.K. (1988). The Influence of the Electronic Structure of Solidson the Anodic Dissolution and Leaching of Semiconducting Sulphide Minerals. *Hydrometallurgy*, 21: 155–190.

Dec, W., Mosiałek, M., Socha, R.P., Jaworska-Kik, M., Simka, W. and Michalska, J. (2016). The effect of sulphate-reducing bacteria biofilm on passivity and development of pitting on 2205 duplex stainless steel. *Electrochimica Acta*, 212: 225–236.

Durmoo, S., Richard, C., Beranger, G. and Moutia, Y. (2008). Biocorrosion of stainless steel grade 304L (SS304L) in sugar cane juice. *Electrochimica Acta,* 54: 74–79.

Fang, H.H.P., Xu L.-C. and Chan K.-Y. (2002). Effects of toxic metals and chemicals on biofilm and Biocorrosion. *Water Research*, 36: 4709–4716.

Fonseca, I.T.E., Feio, M.J., Lino, A.R., Reis, M.A. and Rainha, V.L. (1998). The influence of the media on the corrosion mild steel by *Desulfovibrio desulfuricans* bacteria: an electrochemical study. *Ekctrochimica Acta*, 43 (1-2): 213-222.

Gericke, M., Govender, Y. and Pinches, A. (2010). Tank bioleaching of low-grade chalcopyrite concentrates using redox control. *Hydrometallurgy*, 104: 414–419.

Hansford, G.S. and Vargas, T. (2001). Chemical and electrochemical basis of bioleaching processes. *Hydrometallurgy*, 59: 135–145.

Iken, H., Etcheverry, L., Bergel, A. and Basseguy, R. (2008). Local analysis of oxygen reduction catalysis by scanning vibrating electrode technique: A new approach to the study of biocorrosion. *Electrochimica Acta,* 54: 60–65.

Kinzler, K., Gehrke, T., Telegdi, J. and Sand W. (2003). Bioleaching-a result of interfacial processes caused by extracellular polymeric substances (EPS). *Hydrometallurgy*, 71: 83-88.

Landoulsi, J., Elkirat, K., Richard, C., Feron, D. and Pulvin S. (2008a). Enzymatic Approach in Microbial-Influenced Corrosion: A Review Based on Stainless Steels in Natural Waters. *Environmental Science and Technology*, 42: 2233–2242.

Landoulsi, J., Genet, M.J., Richard, C., El Kirat, K., Rouxhet, P.G. and Pulvin, S. (2008b). Ennoblement of stainless steel in the presence of glucose oxidase: Nature and role of interfacial processes. *Journal of Colloid and Interface Science*, 320: 508–519.

Li, Y., Qian, G., Li, J. and Gerson, A.R. (2015). Chalcopyrite Dissolution at 650 mV and 750 mV in the Presence of Pyrite. *Metals*, 5: 1566–1579.

Li, Y., Wei, Z., Qian, G., Li, J. and Gerson, A.R. (2016). Kinetics and Mechanisms of Chalcopyrite Dissolution at Controlled Redox Potential of 750 mV in Sulfuric Acid Solution. *Minerals*, 6 (83): 1–18.

Liu, H.-c., Xia, J.-l., Nie, Z.-y., Wen W., Yang Y., Ma, C.-y., Zheng, L. and Zhao, Y.-d. (2016). Formation and evolution of secondary minerals during bioleaching of chalcopyrite by thermoacidophilic

Archaea *Acidianus manzaensis*. *Transaction of Nonferrous Metals Sociey of China,* 26: 2485–2494.

Liu, H., Xu, D., Yang, K., Liu, H. and Cheng, Y.F. (2018a). Corrosion of antibacterial Cu-bearing 316L stainless steels in the presence of sulfate reducing bacteria. *Corrosion Science*, 132: 46–55.

Liu, X., Shi, L. and Gu, J.-D. (2018b). Microbial electrocatalysis: Redox mediators responsible for extracellular electron transfer. *Biotechnology Advances*, 36: 1815–1827.

Ma, Y.-l., Liu, H.-c., Xia, J.-l., Nie, Z.-y., Zhu, H.-r., Zhao, Y.-d., Ma, C.-y., Zheng, L., Hong, C.-h. and Wen, W. (2017). Relatedness between catalytic effect of activated carbon and passivation phenomenon during chalcopyrite bioleaching by mixed thermophilic Archaea culture at 65°C. *Transaction of Nonferrous Metals Sociey of China*, 27:, 1374–1384.

Magnin, J.P., Garden, J., Ozil, P., Pathe, C., Picq, G., Poignet, J.C., Schoeffert, S., Vennereau, P. and Vincent, D. (1994). Preparation of porous materials by bacterially enhanced corrosion of Fe in iron-titanium hot-pressed plates. *Materials Science and Engineering*, A189: 165-172.

Maluckov, B.S. (2012a). Bofilms and corrosion of stell. *Hemijski pregled*, 53: 119-123.

Maluckov, B.S. (2012b). Corrosion of steels induced by microorganisms. *Metallurgical and Materials Engineering*, 18: 223-231.

Maluckov, B. (2013). Biocorrosion of copper and their alloys. *Tehnika*, 2: 242-244.

Maluckov, B.S. (2014). Ti-based biomaterials-properties and production. *Optoelectronics and advanced materials- rapid communications*, 8: 545-550.

Miranda, E., Bethencourt, M., Botana, F.J., Cano, M.J., Sa´nchez-Amaya, J.M., Corzo, A., Garcı´a de Lomas J., Fardeau, M.L. and Ollivier, B. (2006). Biocorrosion of carbon steel alloys by an hydrogenotrophic sulfate-reducing bacterium *Desulfovibrio capillatus* isolated from a Mexican oil field separator. *Corrosion Science*, 48: 2417–2431.

Niu, Y., Sun, F., Xu, Y., Cong, Zh. and Wang, E. (2014). Applications of electrochemical techniques in mineral analysis. *Talanta*, 127: 211–218.

Padilla-Viveros, A., Garcia-Ochoa, E. and Alazard, D. (2006). Comparative electrochemical noise study of the corrosion process of carbon steel by the sulfate-reducing bacterium *Desulfovibrio alaskensis* under nutritionally rich and oligotrophic culture conditions. *Electrochimica Acta*, 51: 3841–3847.

Paital, R.S. and Dahotre, N.B. (2009). Calcium phosphate coatings for bio-implant applications: Materials, performance factors, and methodologies. *Materials Science and Engineering*, R66: 1–70.

Rajasekar, A. and Ting, Y.-P. (2010). Microbial corrosion of aluminum 2024 aeronautical alloy by hydrocarbon degrading bacteria *Bacillus cereus* ACE4 and *Serratia Marcescens* ACE2. *Industrial and Engineering ChemistryResearch*, 49: 6054-6061.

Ratner, B.D., Hoffman, A.S., Schoen, F.J. and Lemons, J.E., Biomaterials Science, 2nd ed., Elsevier Academic Press, 2004.

Run-lan, Y., Jian-xi, T., Peng, Y., Jing, S., Xiong- jing, O. and Yun-jie, D. (2008). EPS-contact-leaching mechanism of chalcopyrite concentrates by *A. ferrooxidans. Transaction of Nonferrous Metals Sociey of China*, 18: 1427-1432.

Sand, W. (2003). Microbial life in geothermal waters. *Geothermics,* 32, 655-667.

Sheng, X., Ting, Y.P. and Pehkonen, S.O. (2007). Force measurements of bacterial adhesion on metals using a cell probe atomic force microscope. *Journal of Colloid and Interface Science*, 310: 661–669.

Silva, P.M.P., Lucheta, A.R., Bitencourt, J.A.P., do Carmo, A.L.V., Cuevas, I.P.Ñ., Siqueira, J.O., de Oliveira, G.C. and Alves, J.O. (2019). Covellite (CuS) Production from a Real Acid Mine Drainage Treated with Biogenic H_2S. *Metals*, 9: 206.

Starosvetsky, J., Starosvetsky, D. and Armon, R. (2007). Identification of microbiologically influenced corrosion (MIC) in industrial equipment failures. *Engineering Failure Analysis*, 14: 1500–1511.

Starosvetsky J., Starosvetsky D., Pokroy B., Hilel T. and Armon R. (2008). Electrochemical behaviour of stainless steels in media containing iron-oxidizing bacteria (IOB) by corrosion process modelling, *Corrosion Science*, 50: 540–547.

Tao, J., Qian, L., Yong-bin, Y., Guang-hui, L. and Guan-zhou, Q. (2008). Bio-oxidation of arsenopyrite. *Transaction of Nonferrous Metals Sociey of China,* 18: 1433-1438.

Tshilombo, A.F., Petersen, J. and Dixon, D.G. (2002). The influence of applied potentials and temperature on the electrochemical response of chalcopyrite during bacterial leaching. *Minerals Engineering*, 15: 809–813.

Tu, Z., Wan, J., Guo, C., Fan C., Zhang, T., Lu, G., Reinfelder, J.R. and Dang, Z. (2017). Electrochemical oxidation of pyrite in pH 2 electrolyte. *Electrochimica Acta*, 239: 25–35.

Videla, H.A. (1994). Biofilms and Corrosion Interactions on Stainless Steel in Seawater. *International Biodeterioration and Biodegradation*, 245-257.

Videla, H.A. (2001). Microbially induced corrosion: an updated overview. *International Biodeterioration and Biodegradation*, 48: 176-201.

Wang, W., Wang, J., Xu, H. and Li, X. (2006). Electrochemical techniques used in MIC studies. *Materials and Corrosion*, 57(10), 800–804.

Wang, N., Fang, D., Zheng, G., Liang J. and Zhou L. (2019). A novel approach coupling ferrous iron bio-oxidation and ferric iron chemo-reduction to promote biomineralization in simulated acidic mine drainage. *RSC Advances*, 9: 5083-5090.

Watling, H.R. (2006). The bioleaching of sulphide minerals with emphasison copper sulphides - a review, *Hydrometallurgy,* 84: 81-108.

Watling, H.R., Collinson, D.M., Shiers, D.W., Bryan, C.G. and Watkin, E.L.J. (2013). Effects of pH, temperature and solids loading on microbial community structure during batch culture on a polymetallic ore. *Minerals Engineering*, 48: 68–76.

Wiertz, J.V., Mateo, M. and Escobar, B. (2006). Mechanism of pyrite catalysis of As(III) oxidation in bioleaching solutions at 30°C and 70°C. *Hydrometallurgy,* 83: 35-39.

Xu, C., Zhang, Y., Cheng, G. and Zhu, W. (2008). Pitting corrosion behavior of 316L stainless steel in the media of sulphate-reducing and iron-oxidizing bacteria. *Materials characterization* 59: 245–255.

Yuan, S.J. and Pehkonen, S.O. (2007). Microbiologically influenced corrosion of 304 stainless steel by aerobic *Pseudomonas* NCIMB 2021 bacteria: AFM and XPS study. *Colloids and Surfaces B: Biointerfaces*, 59: 87–99.

Yuan, S.J., Choong, A.M.F. and Pehkonen, S.O. (2007). The influence of the marine aerobic Pseudomonas strain on the corrosion of 70/30 Cu–Ni alloy. *Corrosion Science* 49: 4352–4385.

Zhang, Z., Bi, X., Li, X., Zhao Q. and Chen H. (2018). Schwertmannite: occurrence, properties, synthesis and application in environmental remediation. *RSC Advances,* 8: 33583-33599.

Zhao, X., Duan, J., Hou, B. and Wu, S. (2007). Effect of Sulfate-reducing Bacteria on Corrosion Behavior of Mild Steel in Sea Mud. *Journal of Materials Sciience and Technology*, 23 (3): 324-328.

In: Horizons in World Physics. Volume 302 ISBN: 978-1-53617-180-8
Editor: Albert Reimer

Chapter 2

COATING LIQUID FILMS: A BRIEF REVIEW

Zijing Ding[1,*] and Chun Yang[2]

[1]Department of Applied Mathematics and Theoretical Physics,
University of Cambridge, United Kingdom.
[2]School of Mechanical and Aerospace Engineering,
Nanyang Technological University, Singapore

ABSTRACT

Liquid film flows coating a solid surface have received much attention during last decades due to their vast industrial applications, such as surface protection, lubrication, cooling etc. To understand the dynamics of the film flows, involving instability, pattern formation, rupture, the long-wave models were used based on the fact that the film thickness is much smaller than a typical wave length. In literature studies, past researches focused on two cases: a liquid film flowing down a flat plane and a liquid film coating on a circular cylinder. In this review, we focus on the latter case: a liquid film on a circular cylinder. This flow is essentially unstable because of the famous Plateau-Rayleigh mechanism.

* Corresponding Author's Email: z.ding@damtp.cam.ac.uk.

Therefore, it is interesting to design active control methods to modulate the instability. Here, the influences of wall slippage, thermocapillary force and electric field will be reviewed. Finally, we will suggest several future research directions, including flow control and optimization and machine learning.

1. INTRODUCTION

Over the past few decades, due to the vast applications in coating technology, liquid film flows have received much attention. The film flows are also widely encountered in biological systems, e.g., tear films in human eyes and lubrication films in human lungs. In geophysical systems, film flows, such as gravity currents, ice sheets, are widely attracting scientists' attention. Many experimental and theoretical works have been devoted to two typical flow cases: (1) a liquid film flowing down a flat plane (Oron et al. 1997), and (2) a liquid film flowing on a circular cylinder (Kliakhandler et al. 2001, Craster, & Matar 2006). The latter case is the focus of this review, which has great potential applications in oil recovery and biological flows. The famous Plateau-Rayleigh mechanism modulated with the gravity gives rise to many interesting dynamical phenomena, e.g., organized bead trains and breakup of liquid films (Duprat et al. 2009).

To investigate the complex dynamics in a liquid film coating the exterior surface of a cylinder, most theoretical works have employed a low-dimensional model. The pioneering work was carried out by (Frenkel, 1992), who assumed that the film thickness h is far smaller than the cylinder radius a and used a thin film model to describe the evolution of the interface. Linear stability analysis by Frenkel (1992) showed that the azimuthal curvature component is destabilizing, while the axial curvature component is stabilizing. The nonlinear dynamics was examined by (Kalliadasis, & Chang 1994; Chang 1999; Yu & Hinch 2013), who showed that large droplet can emerge due to the strong nonlinearity. Lister et al. (Lister et al. 2006) showed that the minimal thickness between a lobe and a collar reduces as $t^{-1/2}$ and the minimal thickness between two collars thins as t^{-1}, which indicates that the film does not break up via the sole Plateau-

Rayleigh mechanism. To account for the rupture dynamics, Chen & Hwang (1996) considered the van der Waals attractions, and showed that the liquid film can break up in finite time. The van der Waals attractions, proportional to h^{-3} or h^{-4}, are strong when the film thickness is around 10 −100 *nm*. Very recently, Ding et al. (2019) employed the thin film model and demonstrated that the minimal thickness reduces as $(t_r - t)^{1/5}$ (t_r is the breakup time) if the van der Waals attractions are proportional to h^{-3}. A shortcoming of this model is that it is only valid when the film is thin. But an advantage of the thin film model over the other models is that it is amendable to a three-dimensional case.

Thick film flow, i.e., the film is thicker than the fibre, cannot be analyzed using the thin film model, which exhibits more interesting dynamic phenomena. The experimental study by Kliakhandler et al. (2001) demonstrated that three different flow regimes, labeled "a," "b," "c" can be observed, depending on flow rate. At a high flow rate (~22mm^3/s), long-space separated droplets are propagating at a constant speed, which is referred to as flow regime "a." Flow regime "b" is operated at a moderate flow rate (~11mm^3/s), which is a necklace-like flow structure. In flow regime "c," the flow rate is low (~5mm^3/s). It is an unsteady flow, in which a big droplet is living with several much smaller droplets. The big droplet is consuming these small droplets, while the big droplet itself is unstable in the back region, which breaks up into small droplet again. Kliakhandler et al. proposed a thick film model to investigate the dynamics, in which the full curvature was retained. However, they failed to use this model to predict the dynamics in flow regime "a." They claimed that flow regime "c" can be predicted by the thick film model through numerical simulation. It should be indicated that the thick film model is not self-consistent because of the retained full curvature term. Craster and Matar (2006) revisited this problem and reduced the curvature asymptotically, which gives a consistent description of the problem. The asymptotic model by Craster and Matar (2006) successfully predicted the wave speed of flow regime "a." However, it showed a poor agreement with flow regime "b," which is probably because flow regime "b" is in short wave regime. The unsteady flow regime “c” due to droplet coalescence and breakup in previous studies, however, is compared with a steady traveling wave solution. Very recently, Ding and Willis (2019) pointed out that steady traveling waves cannot be used to describe the flow regime “c” since they do not capture the unsteady dynamics. Instead, they used a relative periodic solution to describe the flow regime “c,” which captures a correct film profile and the unsteady dynamics.

When inertial effects are taken into account, two coupled equations governing the film thickness and the flow rate can be derived, such as the integral-boundary-layer model (Sisoev, Craster, Matar & Gerasimov 2006; Shkadov, Beloglazkin & Gerasimov 2008), the energy integral model (Novbari & Oron 2009) and the weighted-residual model (Ruyer-Quil et al. 2008; Ruyer-Quil & Kalliadasis 2012). The energy integral model and the weighted-residual model are more accurate than the thick film model and the asymptotic model. For instance, a small capillary ripple in front of the big droplet predicted by the asymptotic model is not obvious in experiment, but is eliminated in the weighted-residual model (Ruyer-Quil et al. 2008; Ruyer-Quil & Kalliadasis 2012).

The thin film model, the thick film model, the asymptotic model and the two-coupled-equation models have been applied to many different situations, such as a liquid film flowing down a cylinder with wall slippage, a heated film, electrified film, etc. Below, we will briefly review literatures on wall slippage, thermocapillary effects and electric field on the linear and nonlinear dynamics of the liquid film flowing systems.

2. Governing Equations

We consider a liquid film of mean radius s_0 flowing down a vertical cylinder of radius r_0. The motion of Newtonian liquids is governed by the continuity equation and the momentum equation as below,

$$\nabla \cdot \boldsymbol{u} = 0 \tag{1}$$

$$\rho \frac{D\boldsymbol{u}}{Dt} = -\nabla p + \mu \nabla^2 \boldsymbol{u} + \rho \boldsymbol{g} \tag{2}$$

where $\boldsymbol{u} = ue_r + ve_\theta + we_z$ is the velocity. ρ is the density of the liquid and μ is the dynamical viscosity. $\boldsymbol{g}$ denotes the gravitational acceleration.

At the liquid ring's interface $r = h(z,t)$, the stress balance condition is expressed as:

$$(\mathrm{T}_l - \mathrm{T}_g) \cdot \boldsymbol{n} = -\gamma (\nabla \cdot \boldsymbol{n}) \boldsymbol{n} \tag{3}$$

where T_l or T_g is the stress tensor in the liquid phase and gas phase respectively. γ is the surface tension and $\boldsymbol{n}$ is the interface normal.

Experiments have shown (Kliakhandler et al. 2001; Craster & Matar 2006) that the flow rate is very low, i.e., the Reynolds number $Re = \rho W s_0 / \mu$ is very small, such that the inertia is negligible. When the typical axial length L is much longer than the mean film thickness or the mean film radius, we have $\frac{\partial^2 u}{\partial z^2} \ll \frac{\partial^2 u}{\partial r^2}$ such that the streamwise dissipation is of higher order, which is negligible. In this review, we will focus on small Reynolds number flows and use long-wave models to investigate the dynamic phenomena. However, the streamwise curvature in $-\nabla \cdot \boldsymbol{n}$ cannot be dropped, which is reflected by a WKB analysis or singular perturbation technique (Craster & Matar, 2006).

3. Wall Slippage

In many practical cases, the wall surface is slip: the surface can be rough at the microscale and the superhydrophobic effect would cause a considerable slippery length. The slippery length can be up to 50μm in the case of grooved surfaces (Voronov & Papavassiliou 2008). A Navier-slip boundary condition is widely used to model the wall slippage:

$$\beta \boldsymbol{t} \cdot (\nabla \boldsymbol{u} + \nabla \boldsymbol{u}^T) \cdot \boldsymbol{n} = \boldsymbol{t} \cdot (\boldsymbol{u} - \boldsymbol{u}_w) \tag{4}$$

where $\boldsymbol{t}$ and $\boldsymbol{n}$ are the wall tangent and normal, respectively. This means that the tangential stress is linearly correlated with the tangential velocity difference. $\boldsymbol{u}_w$ is the wall moving speed which is zero in this review β is the Navier slip length in this section.

When applying Eq.(4) to a two dimensional axisymmetric flow, we can have

$$\beta w_r = w. \tag{5}$$

A thick model was derived by Ding & Liu (2011) to investigate the interfacial dynamics

$$s_t + \frac{1}{s}[Req + q\sigma(\frac{s_{zz}}{(1+s_z^2)^{3/2}} - \frac{1}{s(1+s_z^2)^{1/2}})_z]_z = 0 \tag{6}$$

where

$$q = \frac{1}{16}[4s^2 \ln\frac{s}{\alpha} - (\alpha^2 - s^2)(\alpha^2 - 3s^2) + \frac{4\beta(s-\alpha)^2(\alpha+s)^2}{\alpha}] \tag{7}$$

$Re = gs_0^3 / v^2$ is the Reynolds number, $\alpha = r_0 / s_0$ and $\sigma = \frac{\gamma s_0}{\rho v^2}$ is the dimensionless surface tension, in which the mean film radius s_0 is used as the radial length scale. Here s is the liquid film radius which is connected to the film thickness h by $s = \alpha + h$.

When the film thickness is much thinner than the cylinder radius $\alpha \gg h$, we have a thin film model below, governing the evolution of the film thickness

$$h_t + [(\frac{h^3}{3} + \beta h^2)(1 + \frac{1}{Ca}(h_z / \alpha^2 + h_{zzz}))]_z = 0 \tag{8}$$

Which was used by Chao, Ding & Liu (2018) to investigate a uniformly heated film flowing down a cylinder. Here, we use a different time scale $t \rightarrow t / Re$ and $Ca = Re/\sigma$.

Linear stability analysis around $s = 1$ of Eq.(6) gives the following dispersion relation:

$$\lambda = \frac{\sigma q}{1+\alpha}(k^2 - k^4) - ikRe(\ln\frac{1}{\alpha} + \frac{\alpha^2-1}{2} + \frac{\beta}{\alpha}(1-\alpha^2) \tag{9}$$

where k is the wave number, λ is the complex growth rate of disturbance and $i = \sqrt{-1}$.

When $\alpha \rightarrow 1$, the dispersion relation given by the thin film model Eq.(8) is

$$\lambda = \frac{1}{Ca}(1/3 + \beta)k^2(1/\alpha^2 - k^2) - (1+2\beta)ik\,. \tag{10}$$

Linear stability analysis demonstrates that the effective growth rate *Real*(λ) increases as β increases, indicating that the wall slippage is enhancing the Plateau-Rayleigh Mechanism. Physically, with increasing the slip length β, the friction in the system reduces. As viscous friction dissipates kinetic energy of disturbances, it indicates that the wall slippage is destabilizing the interface. The physical mechanism has been verified experimentally by Haefner et al. (2015).

The linear wave speed predicted by the thick film model shows that it is promoted by the wall slippage:

$$c_l = Re(\ln\frac{1}{\alpha} + \frac{\alpha^2-1}{2} + \frac{\beta}{\alpha}(1-\alpha^2))\,. \tag{11}$$

The thin film model shows that the wave speed is increased to

$$c_l = 1 + 2\beta. \tag{12}$$

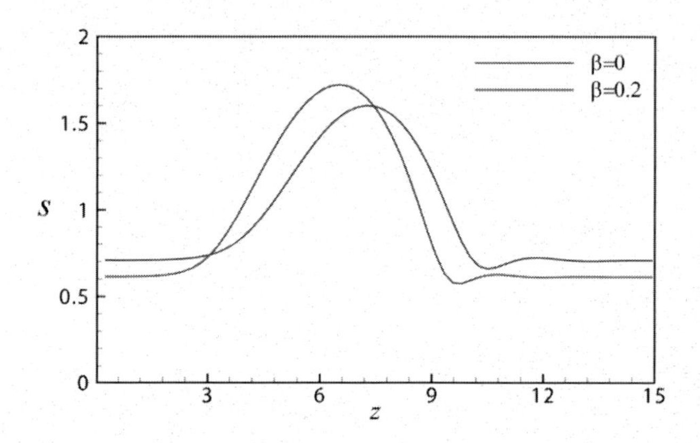

Figure 1. Typical wave profile at $\sigma = 10$, $Re = 1$, $\alpha = 0.5$.

The linear stability analysis of the thick film model shows that the interface is unstable to long-wave disturbances $0 < k < \frac{1}{\alpha+1}$, which is the well-known Plateau-Rayleigh mechanism. When the interface is perturbed by long-wave disturbances, the system could evolve into steady traveling waves. To investigate the traveling wave dynamics, we introduce the transformation:

$$\xi = z - ct, \quad s(z,t) = s(\xi) \tag{13}$$

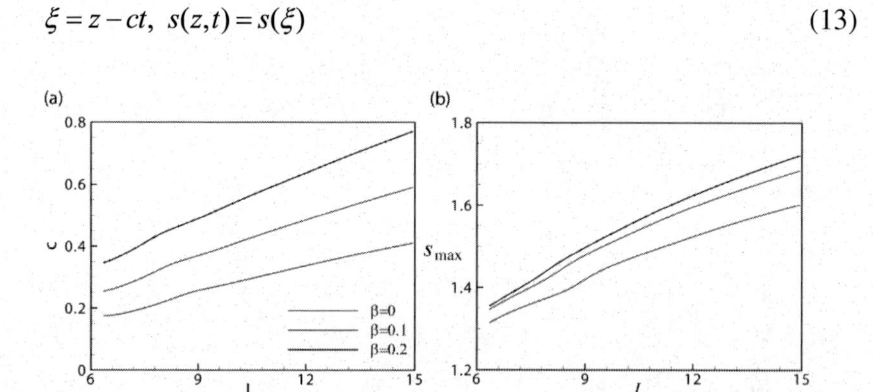

Figure 2. (a) The wave speed c vs. the domain size L. (b) The maximal wave height $s_{\max}$ vs. the domain size.

We used our Newton-Krylov-hookstep code (Ding & Willis 2019) and tracked the single-hump traveling wave, which is born from a Hopf bifurcation at $k_c = 1$. Typical wave profiles computed by the thick film model are shown in Figure 1. The wave speed and height are increased by the wall slippage as demonstrated in Figure 2. When the flow rate is moderate (*Re* = *O*(10)), an integral boundary layer model is more accurate than the thick film model (Ding, Wong, Liu & Liu 2013).

4. THERMOCAPILLARY EFFECT

4.1. Interfacial Instability

In industrial applications, coating of fiber is operated in non-isothermal environments, for example, pulling out a cylinder from a hot liquid bath (Johnson, & Conlisk 1987) or cooling of hot glass fibers (Sweetland & Lienhard 2000). Thermocapillary effect would be significant when the temperature difference between the cylinder surface and the surrounding environment is large. The surface tension is sensitive to the interfacial temperature, which is not uniform when the interface is perturbed. For a normal fluid, its surface tension is linearly dependent on the interface temperature as:

$$\gamma = \gamma_0 + \frac{d\gamma}{dT}(T_i - T_0) \tag{14}$$

where $\frac{d\gamma}{dT} < 0$. The magnitude of the thermocapillary force is characterized by a Marangoni number defined as:

$$Ma = \frac{d\gamma}{dT}\frac{T - T_0}{\rho g s_0^2} \tag{15}$$

An asymptotic model was used by Liu & Liu (2014) to investigate the thermocapillary force on the dynamics:

$$s_t + \frac{1}{s}[(1 + \frac{s_z}{s^2} + \epsilon^2 s_{zzz})(\frac{s^4}{4} ln(s/\alpha) + (3s^2 - \alpha^2)(\alpha^2 - s^2)/16)$$
$$-\epsilon MaT_{iz}(\frac{s^3}{2} ln(s/a) - (s^3 - \alpha^2 s)/4)]_z = 0 \tag{16}$$

where T_i is the interfacial temperature

$$T_i = \frac{1}{1 + Bis \ln(s/\alpha)}. \tag{17}$$

The parameter Bi is the Biot number measuring the resistance ratio of heat conduction to heat convection across the interface. The small parameter $\epsilon = s_0 /L$ can be connected to the Bond number $B_o = \rho g s_0^2 / \gamma$ when the capillary length $L = \gamma / \rho g s_0$ is chosen as the axial length scale.

When $\alpha \to 1$ $(h \ll \alpha)$ and $B_i \ll 1$ (but $MaBi$ remains finite), the asymptotic model reduces to the thin film model (Chao, Ding & Liu 2018; Ding et al. 2018):

$$h_t + h^2 h_z + [\frac{h^3}{3Ca}(h_z + h_{zzz}) + \frac{MaBi}{2}\frac{h^2 h_z}{(1+Bih)^2}]_z = 0 \tag{18}$$

where Ca can be connected to the small parameter by $Ca = \epsilon$.

Linear stability analysis of the thick model (16) gives the following dispersive relation:

$$\lambda + \frac{k^3(k^2\epsilon^2-1)}{16}(-4\,ln(\alpha) - \alpha^4 + 4\alpha^2 - 3)$$
$$+\frac{\epsilon MaBi}{4}\frac{(\alpha^2-1-2\,ln(\alpha))(ln(\alpha)-1)}{(Bi\,ln(a)-1)^2} = 0 \tag{19}$$

As $\alpha \to 1$, we have

$$\lambda = -ik + \frac{1}{3Ca}k^2(1-k^2) + \frac{MaBi}{2(1+Bi)^2}k^2 \tag{20}$$

The dispersive relation (19) and (20) show that the interface is stabilized when Ma < 0 but is destabilized when Ma > 0 (Liu & Liu, 2014; Ding et al. 2018). The physical mechanism for the interfacial instability modified by the thermocapillary force was explained by Ding et al. (2018). Here, we consider a heated film and assume that the interface is perturbed such that the film thickness reduces at some point. The temperature at the trough rises which is higher than its vicinity region. The surface tension at the trough is lowest. As a result, the fluid is driven away from the trough by the surface tension gradient, which enhances the deformation of the interface. Therefore, the instability is enhanced by the Marangoni effect. Numerical simulations showed that the Marangoni effect can modify the types of bound state. When the Marangoni effect is strong, more small droplets are seen in the numerical simulation and oscillatory rather than steady flow state is observed due to the droplet coalescence and breakup processes.

Another interesting question is that: does the Marangoni effect cause the rupture of the film? Very recently, Ding et al. (2019) carried out a similarity analysis of Eq. (18):

$$h = \Delta t^{\gamma} F(\zeta),\ \Delta t = t_r - t,\ \zeta = \frac{z - z_r}{(t_r - t)^{\eta}} \tag{21}$$

where z_r is the rupture location and t_r is the rupture time.

Balancing the time evolution term, the surface tension term and the Marangoni term gives

$$\gamma = -1,\ \eta = -\frac{1}{2} \tag{22}$$

which implies that the rupture time $t_r \to \infty$. The similarity analysis and numerical study indicate that the Marangoni effect would not cause a "true" rupture of the film but accelerates the thinning process. When the evolution time is very long (film at the neck region is very thin), the "rupture" phenomenon is due to numerical errors.

For uniformly heated liquid films discussed above, the Marangoni effect always enhances the Plateau-Rayleigh mechanism. This conclusion, however, does not apply to non-uniformly heated liquid films. When there is a temperature gradient along the cylinder, e.g., heating the fiber on one side and cooling it on the other side, if the temperature gradient can cause a reversal flow along the interface, the thermocapillary effect can inhibit the instability (Dijkstra & Steen 1991; Chen, Abbaschian & Steen 2003; Liu, Ding & Chen 2018).

4.2. Absolute/Convective Instability

Two different flow patterns were observed in experiments when the film is isothermal (Liu, Ding & Zhu 2017). In one pattern, droplets were dispersed in the whole working space, which is caused by an absolute instability. In the other pattern, droplets were only observed in the downstream region, which is caused by a convective instability. For more details of absolute/convective instability, a good review paper was provided by Huerre and Monkewitz (1990). The absolute/convective instability can also be modified by the Marangoni effect (Ding et al. 2018). To determine the instability type, the most amplified wave of zero group velocity is examined:

$$\frac{d\lambda}{dk} = 0 \tag{23}$$

If the real part of the eigenvalue $\lambda > 0$, the instability is absolute. Else, the instability is convective.

Using the thin film model (18), Ding et al. (2018) derived an analytical condition for absolute instability:

$$Ca^{-\frac{2}{3}} + \frac{3MaBi}{2(1+\mathrm{Bi})^2}\mathrm{Ca}^{\frac{1}{3}} > [\frac{9}{4}(-17+7\sqrt{7})]^{\frac{1}{3}} \approx 1.507. \tag{24}$$

A composite Marangoni number $M = \frac{3MaBi}{2(1+Bi)^2}$ can be introduced here. It can be shown that the instability is always absolute if $Ca < 0.54$. Interestingly, the instability is always absolute which is independent of the surface tension as $M > 0.7119$ (Ding et al. 2018). Numerical simulations of the thin film model are in excellent agreement with the linear stability analysis and Figure 5 shows how the flow pattern changes as the Marangoni effect changes.

4.3. Three-Dimensional Instability

A linear stability analysis demonstrated that the heated liquid film can also be unstable subject to an asymmetric mode (Davalos-Orozco & You 2000). To investigate the nonlinear dynamics, Ding and Wong (2017) modified the two-dimensional thin film model to the three-dimensional case:

$$\frac{\partial h}{\partial t} + \frac{\partial}{\partial z}(\frac{h^3}{3}) + \nabla \cdot [\frac{h^3}{3Ca}\nabla(h + \nabla^2 h) + \frac{Mah^2\nabla h}{2(1+Bih)^2}] = 0 \tag{25}$$

where $\nabla = \boldsymbol{e}_\theta \partial/\partial\theta + \boldsymbol{e}_z \partial/\partial z$. It is interesting that the asymmetric mode is always stable when the Marangoni effect is absent. This explains why the axisymmetric flow is observed in experiments. When the liquid film is heated, numerical simulations with different initial conditions demonstrate that the existence of symmetric and asymmetric states as seen in Figure 3. Interestingly, Ding and Wong (2017) found that an asymmetric flow can emerge due to a spontaneous symmetry breaking in their numerical simulation. To explain this phenomenon, linear stability of traveling wave solutions was examined, which showed that small beads are unstable to an asymmetric mode and large beads are unstable to symmetric mode. It indicates that large beads will break up into small beads and small beads may evolve into a non-symmetric state.

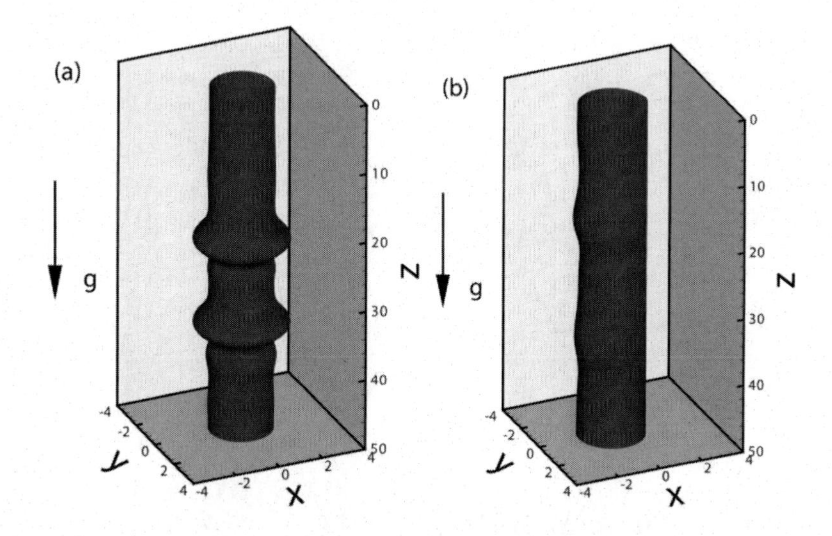

Figure 3. Profiles of the liquid interface. Ma = 2, Bi = 0.1, L_z= 50, ϵ = 0.1. (a) Initial condition: an axisymetric disturbance. (b) Initial condition: non-axisymmetric disturbance. Reproduced from Ding & Wong, Phys. Fluids *29*, 1, 2017, with the permission of AIP Publishing

4.4. Self-Rewetting Films

Very recently, Yu (2018) and Liu, Chen & Wang (2019) investigated self-rewetting liquid films. They showed that the self-wetting liquid film behaves very differently from normal liquid films. For self-rewetting liquids, the surface tension is quadratically dependent on temperature:

$$\gamma = \gamma_0 + \frac{d^2\gamma}{2dT^2}(T_i - T_0)^2, \ \frac{d^2\gamma}{2dT^2} > 0 \qquad (26)$$

The surface tension increases as the interface temperature T_i departs away from the equilibrium state $T = T_0$. When the interface temperature is lower than the reference temperature (the surface tension is minimal at the reference temperature T_0), the thermocapillary effect enhances the Rayleigh-Plateau instability (Yu 2018) and accelerates the droplet moving speed (Liu, Chen & Wang 2019). The conclusion however is opposite when the interface temperature is higher than the reference temperature.

4.5. Fingering Instability

When the fiber/cylinder is partially wetted by liquid, the three-phase (solid-liquid-air) contact line can experience a fingering instability, which was reported by Smolka & SeGall (2011) experimentally and theoretically. Mayo, McCue & Moroney (2013) used a thin film model to revisit this problem, and a three-dimensional simulation of the thin film model showed a nice agreement with experiments. Very recently, Ma (2019) discussed the fingering instability subjected to thermocapillary effect, wherein the fingering instability was enhanced. However, there are still many open questions in these studies. For example, the contact line is modeled as a thin precursor film. The precursor film model, however, loses the physical information of three-phase contact line. Alternatively, the dynamic contact line model was widely used in the study of droplet spreading (Dussan 1979). For a droplet or a gravity current flowing down a vertical cylinder, how does the dynamic contact line model compare with the precursor film model is unknown.

5. Electrified Film

In industrial applications, designing an active control method to modulate the interfacial instability for pattern formations is of particular interest (Papageorgiou, 2018). Using an electric field for this purpose has been successful in many situations, e.g., using an electric field to control pattern formation in a liquid film on a flat plane (Schäffer et al. 2000; Craster & Matar 2005) and droplet production (Conroy et al. 2011; Wang & Papageorgiou 2011). This idea was also transplanted into the control of instability in liquid film flows on vertical fibres (Wray, Matar & Papageorgiou 2012; Wray, Papageorgiou & Matar, 2013(a,b); Ding et al. 2014; Liu, Chen & Ding 2018). The asymptotic model was used to investigate the dynamics of the electrified films (Wray, Matar, & Papageorgiou, 2012; Wray, Papageorgiou, & Matar, 2013; Wray, Papageorgiou, & Matar, 2013; Ding, Xie, Wong, & Liu, 2014; Liu, Chen, & Ding, 2018). When the liquid is a leaky dielectric, the electric field can be used to suppress the Plateau-Rayleigh instability or attract the liquid to the electrode, leading to the touchdown structure (Wray, Matar, & Papageorgiou, 2012; Wray, Papageorgiou, & Matar, 2013; Wray, Papageorgiou, & Matar, 2013), For perfectly conducting liquids, a spike-like structure was observed (Ding et al. 2014; Liu, Chen & Ding 2018), where the liquid interface touches the outer electrode. Self-similarity analysis has shown that the minimal film thickness in touchdown state scales as $(\Delta t)^{1/3}$, and the spike-like structure develops as $(\Delta t)^{1/6}$ ($\Delta t = t_s - t$ with singularity formation time t_s). The interested reader may consult the review paper by Saville (1997) for more details on the dielectric models of liquids. Liquids with high electric conductivities can be modeled as perfect conductors, such as molten liquid metal and electrolytes. A co-axial cylindrical electrode is used to enclose the system such that a radial electric field is built on. The electric field imposes a Maxwell stress on the norm direction of the interface, which is like an attracting force due to the outer electrode (Ding et al. 2014). An asymptotic model was used to investigate the stability and dynamics of the electrified film (Ding et al. 2014, Ding & Willis 2019)

$$s_t + \frac{1}{s}[\frac{1-p_z}{4}(s^4 \ln(\frac{s}{a}) + \frac{(\alpha^2 - s^2)(3s^2 - \alpha^2)}{4})]_z = 0 \tag{27}$$

where $p = -E_b\phi_r^2/2 + 1/s - \epsilon^2 s_{zz}$ and the electrostatic force $\phi_r^2 = s^{-2}[\ln(s/\beta)]^{-2}$ at the leading order approximation. The dimensionless radii of the inner fiber and outer electrode are $\alpha = a/s_0$ and $\beta = b/s_0$. $E_b = \epsilon\varepsilon\phi_0^2/\rho g s_0^3$ is the electric Weber number with ε being the electric permittivity. When the electric field is turned off, Eq. (27) reduces to the asymptotic model by Craster & Matar (2006).

Linear stability analysis of the asymptotic model (27) showed that the instability depends on the gap between the inner fibre and the outer electrode β

$$\lambda = \frac{k^2}{16}\left[\frac{E_b(1 - \ln\beta)}{(\ln\beta)^3} + (1 - \epsilon^2 k^2)\right]\left[-4\ln\alpha - (1-\alpha^2)(3-\alpha^2)\right] + \frac{ik}{2}(2\ln\alpha + 1 - \alpha^2). \tag{28}$$

Physically, the perturbed electric force $\bar{E}q_s'$ is responsible for the interfacial instability. $q_s' = -\varepsilon(1 - \ln(\beta))\delta s/\ln(\beta)^2 + O(\delta s^2)$ is the perturbed surface charge density, where δ_s measures the deformation of the interface. $\bar{E}$ is the electric strength at the basic state, which acts in the opposite direction of r. When the interface is perturbed, $q_s' < 0 (q_s' > 0)$ in the depressed (elevated) region when $\beta < e$. Hence, the electric force is enhancing the deformation of the interface and therefore destabilizing the interface. However, when $\beta > e$, the electric force plays an adverse role, thereby stabilizing the interface. A particular case is $\beta = e$. The interface is unstable due to the surface tension effect, it was found that the electric field has a destabilizing effect on the interface when the perturbation is finite, leading to an oscillatory state or the formation of a singular structure (Ding, Xie, Wong, & Liu, 2014).

Ding & Willis (2019) examined the dynamics of the oscillatory state and demonstrated that this oscillatory state arises from the instability of steady traveling waves. To understand the oscillatory dynamics, they isolated two big droplets in the computational domain, and a relative periodic solution was found to represent the oscillatory state. The relative periodic orbits are closed loops in the energy production and dissipation plane (Ding & Willis, 2019). Additionally, the steady traveling waves are fixed points in the production vs. dissipation plane, which cannot represent an unsteady system.

To identify the solution regimes of steady traveling waves, relative periodic orbits, and spike-like singular structure, Ding & Willis (2019) sought for relative periodic solutions in the parametric space (L, E_b). The regime of the relative periodic solutions is sandwiched by the steady traveling solution regime and the spike-like solution regime The phase diagram shows that steady traveling waves survive when the electric field is weak. While increasing the electric field strength, oscillatory state, which is represented by a temporally recurrent solution (or relative periodic orbit arises. Spike-like singular structure finally emerges when the electric field is strong.

Three-dimensional instability can arise due to the applied electric field (see Wray, Papageorgiou & Matar 2013a; Liu, Chen & Ding 2018). When the film is thin, the three-dimensional instability could dominate the two-dimensional axisymmetric instability (Liu, Chen & Ding 2018), which can develop into nonlinear asymmetric state. When the film is thick, the axisymmetric instability dominates the system unless the electric field is strong enough. However, the linear stability analysis is performed on a uniform film, namely the base state $s = 1$. This implies that the electric field is applied before the formation of droplets by Rayleigh-Plateau mechanism. In this situation, three-dimensional asymmetric steady could arise. However, when the electric field is applied after droplet formation by the Rayleigh-Plateau mechanism, it is unknown whether or not the electric field can trigger an asymmetric state. This should be examined in future theoretical and experimental studies.

6. PERSPECTIVES

Below, we shall propose two potential research directions in thin film research.

6.1. Flow Control and Optimization

We have reviewed the influences of wall slippage, thermocapillary effect, and electric field on the stability of the liquid film. For instability control, all these methods are not very flexible since the flow rates in real applications can vary with time and noises are not as simple as a harmonic wave. It is therefore important to design suitable feedback control strategy in real applications. Very recently, Tomlin et al. (2019) proposed an optimal control method to prevent the exponential growth of transverse waves for the case of a hanging film. The blowing and suction actuators are used for the controlling purpose. This results in an interesting variational problem. Although optimization problem has been investigated greatly in transitional turbulent flows (Kerswell 2018), there are less works on optimization in thin liquid film flows. Therefore, it would be interesting to design versatile and real-time control techniques in thin film flows, e.g., optimizing the electric force.

6.2. Machine Learning

It should be pointed out that the problem reviewed here is essentially nonlinear. For fast control purpose, a linear input-output system is more appealing. Very recently, Koopman analysis has been a powerful candidate to transform a nonlinear system into a linear system using the Koopman mode decomposition (Mezić 2013). A key step in the Koopman analysis is to identify the Koopman modes, which will be very difficult for a generic nonlinear system. Luckily, machine learning is very good at finding the Koopman modes through a dynamic mode decomposition (Brunton 2019).

Therefore, once we identify the Koopman modes, we can manipulate the dominant modes and control the system using a reduced order model, which is much faster than solving a nonlinear equation.

Another interesting issue is that the connection between the reported periodic solutions by Ding & Willis (2019) and the chaotic motions in the liquid film flows is unknown. It is equivalent to ask if a chaotic attractor, which shadows the periodic orbits, can be represented by a superposition of the periodic orbits. This has been a long-standing challenge in turbulence research (Chandler & Kerswell 2013, Budanur et al. 2017). There are two key issues to be addressed: (1) how many periodic solutions are enough and (2) what is the superposition? We can use a bifurcation analysis to identify enough periodic solutions in the film flow, which is impossible in turbulent pipe flow nor in turbulent channel flow. The superposition may be found using a machine learning technique. The answer to question—if chaotic attractor can be represented by periodic solutions, will shed light into the turbulent research, which however is much easier to solve than attacking the turbulent flows directly.

REFERENCES

Brunton S. (2019) *Notes on Koopman Operator Theory*. EPSRC summer school. Cambridge.

Budanur, N., Short, K., Farazmand, M., & Willis, A. (2017). Relative periodic orbits form the backbone of turbulent pipe flow. *J. Fluid Mech, 833*, 274.

Chandler, G., & Kerswell, R. (2013). Invariant recurrent solutions embedded in a turbulent twodimensional Kolmogorov flow. *J. Fluid Mech, 722*, 554.

Chang, C. H., & Demekhin, E. (1999). Mechanism for drop formation on a coated vertical fibre. *J. Fluid Mech, 380*, 233.

Chao Y., Ding, Z., & Liu, R. (2018). Dynamics of thin liquid films flowing down the uniformly heated/cooled cylinder with wall slippage. *Chemical Engineering Science, 175*, 354.

Chen, J., & Hwang, C. (1996). Nonlinear rupture theory of a thin liquid film on a cylinder. *J. Colloid and Interface Sci, 182*, 564.

Chen, Y., Abbaschian, R., & Steen, P. (2003). Thermocapillary suppression of the Plateau-Rayleigh instability: a model for long encapsulated liquid zones. *J. Fluid Mech, 485*, 97.

Conroy, D., Matar, O., Craster, R., & Papageorgiou, D. (2011). Breakup of an electrified perfectly conducting, viscous thread in an AC field. *Phys. Rev, E 83*, 066314.

Craster, R., & Matar, O. (2005). Electrically induced pattern formation in thin leaky dielectric films. Phys. *Fluids 17*(3), 032104.

Craster, R., & Matar, O. (2006). On viscous beads flowing down a vertical fibre. *J. Fluid Mech, 553*, 85.

Davalos-Orozco, L., & You, X. (2000). Three-dimensional instability of a liquid flowing down a heated vertical cylinder. *Phys. Fluids, 12*, 2198.

Dijkstra, H., & Steen, P. (1991). Thermocapillary stabilization of the capillary breakup of an annular film of liquid. *J. Fluid Mech, 229*, 205.

Ding, Z., & Liu, Q. (2011). Stability of liquid films on a porous vertical cylinder. Phys. *Rev. E 84*, 046307.

Ding, Z., & Willis, A. (2019). Relative periodic solutions Relative periodic solutions in conducting liquid films flowing down vertical fibres. *J. Fluid Mech, 873*, 835.

Ding, Z., & Wong, T. N. (2017). Three-dimensional dynamics of thin liquid films on vertical cylinders with Marangoni effect. *Phys. Fluids, 29*, 1.

Ding, Z., Liu, R., Wong, T., & Yang, C. (2018). Absolute instability induced by Marangoni effect in thin liquid film flows on vertical cylindrical surfaces. *Chemical Engineering Science, 177*, 261.

Ding, Z., Liu, Z., Liu, R., & Yang, C. (2019). Breakup of ultra-thin liquid films on vertical fiber enhanced by Marangoni effect, *Chemical Engineering Science, 199*, 342.

Ding, Z., Wong, T. N., Liu, R., & Liu, Q. (2013). Viscous liquid films on a porous vertical cylinder: Dy-namics and stability. *Phys. Fluids, 25*, 064101.

Ding, Z., Xie, J., Wong, T. N., & Liu, R. (2014). Dynamics of liquid films on vertical fibres in a radial electric field. *J. Fluid Mech, 752*, 66.

Duprat, C., Giorgiutti-Dauphin'e, F., Tseluiko, D., Saprykin, S., & Kalliadasis, S. (2009). Liquid film coating a fiber as a model system for the formation of bound states in active dispersivedissipative noninear media. *Phys. Rev. Lett, 103*, 234501.

Dussan E. B. (1979). Spreading of liquids on solid surfaces: static and dynamic contact lines. *Annu. Rev. Fluid Mech,* 11.

Frenkel, A. (1992). Nonlinear thoery of strongly undulating thin films flowing down vertical cylinders. *Europhys. Lett, 18*, 583.

Haefner S. et al. (2015) Influence of slip on the Plateau–Rayleigh instability on a fibre. *Nature Communications*. 7409.

Huerre, P. & Monkewitz, P. (1990). Local and global instabilities in spatially developing flows. *Annu. Rev. Fluid Mech, 22*, 473.

Johnson, R., & Conlisk, A. (1987). Laminar-film condensation/ evaporation on a vertically fluted surface. *J. Fluid Mech, 184*, 245.

Kalliadasis, S., & Chang, C. H. (1994). Drop formation during coating of vertical fibres. *J. Fluid Mech, 261*, 135.

Kerswell R. (2018) Nonlinear nonmodal stability theory. *Ann. Rev. Fluid* Mech. 50, 319.

Kliakhandler, I., Davis, S., & Bankhoff, S. (2001). Viscous beads on vertical fibre. *J. Fluid Mech, 429*, 381.

Lister, J., Rallison, J., King, A., Cummings, L., & Jensen, O. (2006). Capillary drainage of an annular film: the dynamics of collars and lobes. *J. Fluid Mech, 552*, 311-343.

Liu R., & Liu Q. (2014). Thermocapillary effect on the dynamics of viscous beads on vertical fiber. *Phys. Rev, E 90*, 033005.

Liu R., Chen X. & Wang X. (2019) The effect of thermocapillarity on the dynamics of an exterior coating film flow down a fibre subject to an axial temperature gradient. *Int. J. Heat Mass Trans, 123*, 718.

Liu, R., Chen, X. & Ding, Z. (2018). Absolute and convective instabilities of a film flow down a vertical fiber subjected to a radial electric field. *Phys. Rev. E, 97*, 013109.

Liu, R., Ding, Z. & Zhu, Z. (2017). Thermocapillary effect on the absolute and convective instabilities of film flows down a fibre. *Int. J. Heat Mass Trans, 112*, 918.

Liu, R., Ding, Z., & Chen, X. (2018). The effect of thermocapillarity on the dynamics of an exterior coating film flow down a fibre subject to an axial temperature gradient. *Int. J. Heat and Mass Trans, 123*, 718.

Ma, C. (2019). Fingering instability analysis for thin gravity-driven films flowing down a uniformly heated/cooled cylinder. *Int. J Heat Mass Trans., 136*, 719.

Mayo, L., McCue, S., & Moroney, T. (2013). Gravity-driven fingering simulations for a thin liquid film flowing down the outside of a vertical cylinder. *Phys. Rev, E, 87*, 053018.

Mezić, I. (2013) Analysis of Fluid Flows via Spectral Properties of the Koopman Operator. *Ann. Rev. Fluid Mech.* 45, 357.

Novbari E. & Oron A. (2009) Energy integral method model for the nonlinear dynamics of an axisymmetric thin liquid film falling on a vertical cylinder. Phys. Fluids. 21, 062107.

Oron A., Davis, S. H. & Bankoff, S. G. (1997) Long-scale evolution of thin liquid films. *Rev. Mod. Phys*. 69, 931.

Papageorgiou, D. (2018). Film flows in the presence of electric fields. *Annu. Rev. Fluid Mech, 51*, 155.

Ruyer-Quil, C. & Kalliadasis, S. (2012). Wavy regimes of film flow down a fiber. *Phys. Rev, E 85*, 046302.

Ruyer-Quil, C., Treveleyan, P., Giorgiutti-Dauphin'e, F., & Kalliadasis, S. (2008). Modeling film flows down a fibre. *J. Fluid Mech, 603*, 431.

Saville, D. A. (1997). Electrohydrodynamics: the Taylor-Melcher leaky dielectric model. Annu. Rev. *Fluid Mech, 29*, 27.

Schäffer, E., Thurn-Albrecht, T., Russell, T., & Steiner, U. (2000). Electrically induced structure formation and pattern transfer. *Nature 403*, 874.

Shkadov, V. Ya., Beloglazkin, A. N., & Gerasimov, S. V. (2008). Solitary waves in a viscous liquid film flowing down a thin vertical cylinder. *Mosc. Univ. Mech. Bull, 63*, 122.

Sisoev, G. M., Craster, R. V., Matar, O., & Gerasimov, S. (2006). Film flow down a fibre at moderate flow rates. *Chem. Eng. Sci, 61*, 7279.

Smolka, L., & SeGall, M. (2011). Fingering instability down the outside of a vertical cylinder. *Phys. Fluids 23*, 092103.

Sweetland, M., & Lienhard, V. J. (2000). Evaporative cooling of continuously drawn glass fibers by water sprays. *Int. J. Heat Mass Trans, 43*, 777.

Tomlin R., Gomes S. Pavliotis G. & Papageorgiou D. (2019) Optimal control of thin liquid films and transverse mode effects. *SIAM J. Appl. Dynamical Systems*. 18, 117.

Voronov, R., & Papavassiliou, D. (2008). Review of a fluid slip over superhydrophobic surfaces and its dependence on the contact angle. *Ind. Eng. Chem. Res, 47*, 2455.

Wang, Q. & Papageorgiou, D. (2011). Dynamics of a viscous thread surrounded by another viscous fluid in a cylindrical tube under the action of a radial electric field: breakup and touchdown singularities. *J. Fluid Mech, 683*, 27-56.

Wray, A., Matar, O., & Papageorgiou, D. (2012). Nonlinear waves in electrified viscous film flow down a vertical cylinder. *IMA J. Appl. Math, 77*, 430-440.

Wray, A., Papageorgiou, D. & Matar, O. (2013a). Electrified coating flows on vertical fibres: enhance- ment or suppression of interfacial dynamics. *J. Fluid Mech, 735*, 427-456.

Wray, A., Papageorgiou, D., & Matar, O. (2013b). Electrostatically controlled large-amplitude, non- axisymmetric waves in thin film flows down a cylinder. *J. Fluid Mech, 736*, R2.

Yu, L., & Hinch, J. (2013). The velocity of largeviscous drops falling on a coated verticalfibre. *J. Fluid Mech, 737*, 232.

Yu, Z. (2018). Thermocapillary instability of self-rewetting films on vertical fibers. *Phys. Fluids 30*, 082104.

In: Horizons in World Physics. Volume 302 ISBN: 978-1-53617-180-8
Editor: Albert Reimer

Chapter 3

ANOMALOUS SECOND HARMONIC GENERATION DURING FEMTOSECOND SUPERRADIANT EMISSION FROM SEMICONDUCTOR LASER STRUCTURES

Peter P. Vasil'ev[1,2,*], Richard V. Penty[1] and Ian H. White[1]
[1]Centre for Photonic Systems, University of Cambridge, Cambridge, UK
[2]P.N. Lebedev Physical Institute, Moscow, Russia

ABSTRACT

The chapter presents theoretical and experimental studies of the generation of ultra-bright internal second harmonic during femtosecond superradiant emission in multiple section semiconductor laser structures. Experimentally measured conversion efficiencies are by 1-2 orders of

[*] Corresponding Author's E-mail: pv261@cam.ac.uk.

magnitude greater than those predicted by a standard SHG theory. To explain this fact, a model based on one-dimensional nonlinear Maxwell curl equations without taking into consideration the slowly-varying envelope approximation has been developed. The proposed model explains well all experimental data available to date. We show that the unique features of the superradiant emission in semiconductor media are responsible for the observed abnormal SHG. In particular, it has been demonstrated that strong transient periodic modulation of e-h density and refraction index dramatically affects the process of superradiance in semiconductor media. The presence of coherent non-equilibrium carrier density and refractive index gratings can explain the observed nonlinear effects, including the ultra-strong internal second harmonic generation and superluminal optical pulse propagation. A periodic grating rephases the nonlinear polarization and the generated electromagnetic waves. Strong periodic $\lambda/2$ modulation of the e-h density and corresponding modulation of the refractive index can result in quasi-phase matching conditions in a similar manner as in periodically poled materials for harmonics generation.

Keywords: second harmonic, femtosecond pulses, superradiance, semiconductor lasers

INTRODUCTION

Superradiance (SR) has been extensively studied for a long period of time since the first publication by Dicke in 1954 [1]. SR pulses have been observed in many types of media including gaseous, solid-state, polymer, and bulk, quantum-well, and quantum dot semiconductor devices [2–7]. Due to ultrafast characteristic times of electrons and holes in semiconductors, observable durations of SR pulses are in picoseconds and femtosecond ranges [2, 4]. SR emission generated by semiconductor structures possesses a number of unique features, including ultrastrong internal second harmonic generation (SHG), superluminal pulse propagation [8], and greater coherence than lasing [9]. We have previously reported the generation of high-power femtosecond SR pulses from a number of semiconductor multiple contact laser devices at room temperature [2, 4–9]. The propagation of powerful femtosecond pulses in those devices occurs under the influence of various types of nonlinear interactions of the electromagnetic field of the pulses with the intracavity amplifying/absorbing semiconductor medium. The medium not only amplifies and shapes pulses but experiences strong back impact from them which modifies some properties of the medium. It has been recently demonstrated [8] that a characteristic feature of SR in semiconductors is strong coherent population $\lambda/2$ gratings burned by counter-propagating femtosecond SR pulses in the e-h ensemble density. Due to a strong dependence of the refractive index on the carrier density in semiconductor materials, population gratings cause strong refractive index gratings which affect the propagation of SR pulses through the semiconductor.

In addition, it is well-known that semiconductor materials exhibit strong quadratic nonlinearity and efficient SHG. A number of observations of internal SHG generation in GaAs and InGaAsP semiconductor lasers were reported in the 1960s and 1980s [10–12]. Due to a very strong absorption coefficient γ exceeding 10^5 cm^{-1} at the SH wavelength in active layers of the lasers, the SHG output conversion efficiency from the lasers is extremely low. It has been established in early studies that the observed SHG power comes from the vicinity of the chip facet. The effective SH generation length is less or around the emission wavelength in depth. Since typical peak powers of SR pulses are much greater than these of lasing from the same devices, it is clear that internal SH intensities should be much greater when femtosecond SR pulses are generated.

Indeed, we have observed bright blue/violet emission from infrared GaAs/AlGaAs lasers operating under SR [13]. Furthermore, the measured intensity of SH and the efficiency of the frequency doubling proved to be by 1-2 orders of magnitude larger than that suggested by the theory of SHG and measured experimentally [10-12]. The phenomenon of the ultra-bright internal SHG is important from both a fundamental point of view and applications. On one hand, it provides additional evidence that SR is intrinsically different from lasing. On the other hand, it can be exploited for an efficient frequency doubling when other methods fail.

This Chapter presents a further study of the influence of both population coherent grating and refractive index ones on the internal SH generation during SR pulse emission. By comparing the results of modelling with the experimental data observed previously [2, 4–9, 13], one will clearly see that the formation of transient coherent e-h gratings can explain many features of SR, including unusually strong internal second harmonic generation and superluminal pulse propagation. The modelling shows that the periodic modulation of the nonlinear susceptibility of the medium during the SR pulse generation, caused by population and index gratings, is the most probable physical reason for a significant increase in the internal SHG efficiency. The model explains also why the efficiency of the internal SHG is so different for standard lasing and SR which is observed experimentally.

SUPERRADIANCE IN BRIEF

Superradiance is virtually the cooperative radiative decay of an initially inverted ensemble of quantum oscillators. SR is similar to such quantum optics effects as photon echo, self-induced transparency, optical nutation, and optical adiabatic following. The characteristic features of the SR emission include the quadratic dependence of the intensity on the number of the inverted emitters, highly anisotropic emission pattern, temporal and spatial coherence, after-pulse ringing attributed to coherent Rabi-type oscillations and fluctuations of the parameters of the SR pulses (shape, duration, amplitude and time delay) determined by quantum-mechanical uncertainties both in the optical field and the medium [4]. One of the fundamental features of the SR is the mutual phasing of emitters without which the generation of the SR is not practically possible [14]. SR differs from both spontaneous emission and lasing. Figure 1 provides a simple illustration. Consider a system of two-level emitters at different levels of the resonant pump. If the pump is weak, the ensemble emits spontaneously. The power is low and the radiative emission pattern is more or less uniform without any specific direction. Placing the sample into an optical cavity can result in achieving the laser action at certain parameters of the cavity and the pump. Here, the output power is much greater as compared to the previous case. The emitted optical pulse often exhibits so-called relaxation oscillations and the output spatial pattern locates along the cavity axis.

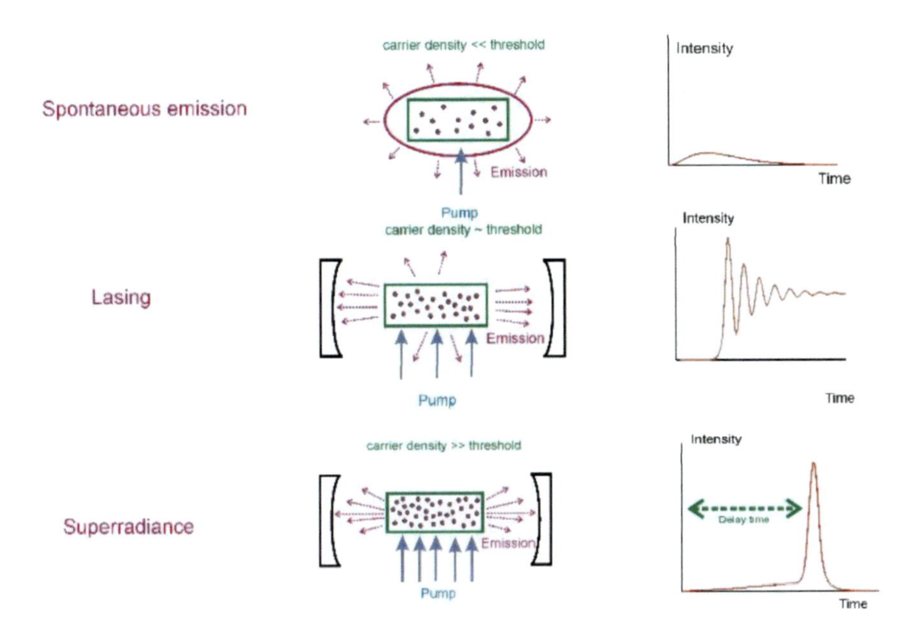

Figure 1. A schematic of spontaneous and laser emission and SR in an ensemble of two-level emitters.

At high density and stronger pumping spontaneous phase-locking of the emitter dipoles can appear throughout the medium. A collective mode, macroscopic polarisation and order appear in the system which can be defined by the build-up of correlation between the dipoles belonging to different emitters. The dipole-dipole correlations eventually emerge through a kind of symmetry breaking phenomenon.

The emission pattern, in this case, is highly nonuniform, the overwhelming part of the intensity being generated along the axis of the sample in both directions. In the time domain, there is a well-defined "incubation" period after which a short giant pulse is emitted. The number of emitters, the spontaneous lifetime, and the geometry of the sample determines the delay time, peak power of the pulse, and its duration.

The self-ordering in the system can only happen at large enough density of the emitters [4, 14]. A simple increase of the pump is not enough for achieving larger densities than the critical value. As the pumping increases, the optical gain in the system also increases and lasing occurs at the threshold density. The onset of lasing clamps the density at the threshold and a further increase of the pump does not result in the density increase [4]. In semiconductor active media there is a simple solution to this problem. The optical gain/absorption in a semiconductor laser structure can be controlled by its splitting in a number of sections and the application of different currents/voltages on them. In particular, lasing in a three-section device with two amplifying sections and one absorbing one can be frustrated by a strong reverse bias on the absorber. The reverse bias allows for much harder pumping of the gain sections and achieving electron-hole densities in the active region by a few times greater than typical lasing values. This technique has been effectively used for the generation of SR femtosecond pulses from semiconductor devices at room temperature for a long time [2, 4, 13]. As mentioned in the Introduction, due to specific properties of semiconductor media SR emission from semiconductor laser devices exhibits a number of unique features, ultrastrong internal SHG being one of them. Let us discuss this effect in detail.

EXPERIMENT

A large variety of multiple section GaAs/AlGaAs bulk and quantum-well laser structures capable for the generation of SR emission were studied. In general, they had gain and saturable absorber sections of different geometries with different gain/absorber ratios and total cavity lengths. The devices under test were described in detail in our previous publications (see, for instance [2, 8, 13]). The composition of the GaAs/AlGaAs heterostructures varied a lot resulting in a broad range of the operating wavelengths between 820 and 890 nm. All devices can operate in c.w. lasing mode, gain/Q-switching or SR regimes depending on the driving conditions. The blue/violet internal SH was observed from all samples along with the fundamental infrared wave. Figure 2 presents a microscopic view of two experimental samples having tapered (a) and straight (b) waveguides. A strong reverse bias of the central absorber section allows for preventing the onset of lasing in the sample for a few nanoseconds and achieving much greater e-h densities than the threshold lasing density. At the appropriate driving conditions [4] SR phase transition occurs in the system and giant superradiant 200-400 femtosecond long pulses with peak powers in the range of 120-200 W are emitted. It has been observed that quite bright blue/violet emission is generated in addition to infrared SR pulses [13]. Figure 3 illustrates a typical SHG beam at around 426 nm which is emitted from the sample in the same direction as the collimated fundamental SR harmonic at 852 nm.

The square blue filter in Figure 3 blocks the infrared beam and prevents overexposure of the CCD detection camera by the fundamental harmonic. The laser was driven by 9 ns long current pulses with an amplitude of around 680 mA at the repetition rate of 15 MHz. The reverse d.c. bias on the absorber section was – 6.4 V. A single femtosecond pulse or bursts of SR pulses were generated per each current pulse. The number of SR pulses was controlled by the current pulse amplitude and/or by the reverse bias. The maximum detected average power of the internal second harmonic was just over 0.2 μW at the fundamental average power of about 1.9 mW.

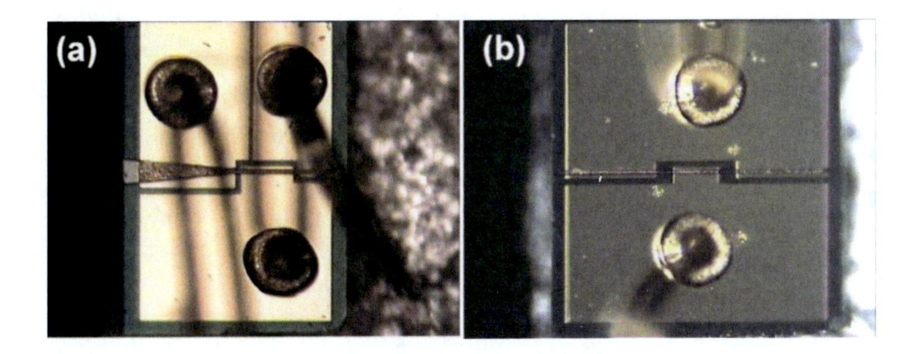

Figure 2. Top view of two experimental samples under test.

Figure 3. The internal SHG beam (the blue horizontal line) as seen using a CCD camera.

The measured SHG power varied a lot among the samples under test depending on parameters of the samples and driving conditions. The internal SH was obviously dependent on the number of SR pulses emitted per each current pulse. Figure 4 presents the measured SH average power versus reverse bias V for 3 samples. The SH power is negligible at reverse biases below around V = 1.0 - 1.5 V. This range corresponds to normal lasing in the devices, the optical output being a nanosecond pulse with weak relaxation oscillations. SR pulses are generated at larger V [2, 4].

In general, the peak power and the number of SR pulses rise first at larger V and the SHG intensity increases sharply. At certain V the number of SR pulses starts to drop rapidly and the SHG average power decreases. Similar behaviour has been observed for all samples. Table 1 gives the results of the measurements of the intensity of the first and second harmonics outside the samples. The last column presents some parameters of the devices, including the cavity length, type of the waveguide (straight or tapered) and its width. The conversion efficiency coefficient η (1/Watt) is determined by the relation of $P_{2\omega} = \eta P_{\omega}^2$ [11, 13].

The values of η for SR pulses measured experimentally by more than the order of magnitude greater than those measured under standard lasing conditions and previously published in [10–12]. Indeed, the published values of η vary between 2.5 x 10^{-8} and 2.9 x 10^{-9} W^{-1}, whereas η in Table 1 lies in the range of 4.7 x 10^{-6} and 1.1 x 10^{-7} W^{-1}. It should be noted that the values of η in Table 1 were calculated for the peak powers of the fundamental and the second harmonic femtosecond pulses outside the laser cavities.

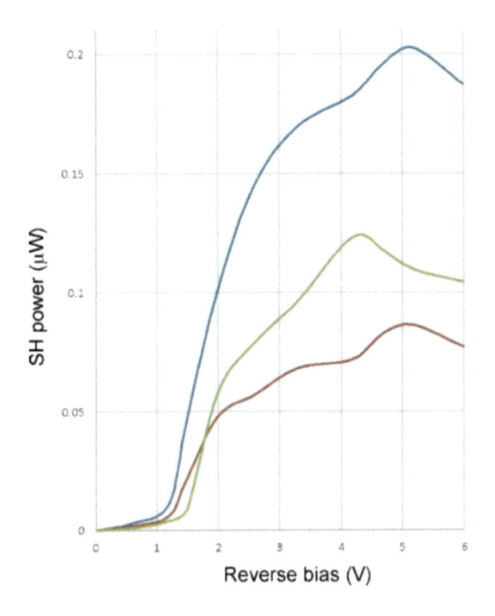

Figure 4. Internal SH intensity versus reverse bias for 3 samples.

Table 1. Internal SH experimental data

Sample	Average P_{ω}	Average $P_{2\omega}$	η(1/W)	Description
H2-51	1.92 mW	193 nW	4.7 x 10^{-6}	250 mm long, 5 mm straight
H2-31	9.2 mW	104 nW	1.1 x 10^{-7}	350 mm long, 5 to 30 mm taper
H2-33	9.7 mW	130 nW	1.3 x 10^{-7}	350 mm long, 5 to 30 mm taper
H2-52	2.3 mW	36 nW	6.2 x 10^{-7}	250 mm long, 5 mm straight
H2-63	3.4 mW	43 nW	3.4 x 10^{-7}	350 mm long, 5 mm straight
H2-112	3.2 mW	41 nW	3.6 x 10^{-7}	450 mm long, 5 mm straight
H2-131	4.1 mW	56 nW	3.0 x 10^{-7}	450 mm long, 5 mm straight
H2-132	7.2 mW	87 nW	1.5 x 10^{-7}	450 mm long, 5 mm straight
Ref [10]	20 W	10 mW	2.5 x 10^{-8}	bulk GaAs, c.w.
Ref [11]	30 mW	2.6 pW	2.9 x 10^{-9}	bulk GaAs/AlGaAs, c.w.
Ref [12]	3.4 mW	1 pW	8.0 x 10^{-8}	QW InGaAsP, c.w.

The internal SH power and SHG conversion efficiencies published in [10–12] were measured outside the cavities for c.w. or quasi-c.w. laser emission. In this case, the peak and average powers were the same. That is why the comparison of the conversion efficiencies is correct since values of peak powers are used for all cases. Strictly speaking, values of the fundamental and second harmonics inside the cavities should be used for the estimation of η. However, the ratio of the second and first harmonic powers inside and outside the cavity is determined by the ratio of the corresponding refractive indices at the two frequencies. The difference of the refractive indices leads to a correction of the calculated η by less than 15-20% for the values in Table 1 as well as calculated in [10–12].

The recorded intensity of SHG and the efficiency of the frequency doubling are by 1-2 orders of magnitude larger than that suggested by conventional theory of SHG in bulk materials. One of the most possible explanations of this discrepancy is the presence of a weak refraction index periodicity of the active medium caused by transient population and index gratings which exist during SR emission generation [8, 13]. We demonstrate now that this fact can be confirmed by a theoretical model which takes into account both population and refractive index transient gratings.

Theoretical Model

SR has been extensively studied theoretically in different media for many years [14–16]. Most of the models are based on solutions of semi-classical or quantum mechanical Maxwell-Bloch equations. Because of the complexity of this treatment, the associated optical field equations have been traditionally simplified using several approaches, including the slowly-varying envelope approximation (SVEA) and rotating wave approximation (RWA). Our previous models of the SR generation in semiconductors [6, 17, 18] were also based on those approaches. It described reasonably well the basic properties of SR pulses. However, the approximations associated with these methods limit their range of applicability when ultrashort pulse phenomena and inhomogeneous media with a fine microstructure are considered.

Finite-difference time-domain (FDTD) solutions of the full-wave vectorial Maxwell–Bloch equations have been previously used for the investigation of nonlinear gain spatio-temporal dynamics of pulse propagation in active optical waveguides and semiconductor microcavities [19–21]. The interaction of the electromagnetic field with transient population gratings can result in significant deviations of experimentally observed SR pulses from results of models which treat the semiconductor as a spatially uniform medium. A model based on full Maxwell equations without the SVEA simplification can consistently take into account all features of the interaction of femtosecond SR pulses with the e-h system having a fine (λ/2) periodic spatial structure. This approach does not require splitting the optical field into counter-propagating waves as well as dividing the carrier density into continuous and spatially oscillating components as has been done in our previous model [8].

Assuming for simplicity that the medium is nonpermeable and isotropic, Maxwell's curl equations in one dimension are [22]

$$\frac{\partial H_y}{\partial t} = -\frac{\partial E_x}{\partial z}, \tag{1}$$

$$\frac{\partial D_x}{\partial t} = -\frac{\partial H_y}{\partial z} \tag{2}$$

$$E_x = \frac{D_x - (P_x^l + P_x^{nl})}{\varepsilon(z)} \tag{3}$$

where ε(z) is the relative permittivity of the medium and the polarization P_x consists of the linear and nonlinear terms

$$P_x = P_x^l + P_x^{nl} = \chi^{(1)} E_x + \chi^{(2)} E_x^2 \tag{4}$$

We treat the semiconductor medium as a two-level system. We consider that as a reasonable assumption because, as we have demonstrated before [4, 23], during the early stages of the development of SR pulses almost all electrons and hole condense in the phase space at the bandgap energy. As opposed to the case of lasing and spontaneous emission when e-h pairs are located over a broad energy range within the conductive and valence bands, SR pulses originate from the recombination of e-h pairs located within two narrow energy regions at the bottom of the conduction band (for the electrons) and at the top of the valence band (for the holes) [23]. There is negligible emission from upper energy levels, and so the density matrix formalism for a two-level system is valid. The Bloch equations for the elements of the density matrix are [19]

$$\frac{\partial u}{\partial t} = \omega_0 v - \frac{u}{T_2}, \tag{5}$$

$$\frac{\partial v}{\partial t} = -\omega_0 u + 2\frac{d}{\hbar} E_x w - \frac{v}{T_2}, \tag{6}$$

$$\frac{\partial w}{\partial t} = -2\frac{d}{\hbar} E_x v - \frac{w - w_0}{T_1} \tag{7}$$

The components u and v represent, respectively, the dispersive or in-phase and the absorptive or in-quadrature polarization components associated with the dipole transition. The polarization of the two-level medium is determined by the relation $\mathrm{Px(t)} = -\mathrm{ndu(t)}$, where d is the transition dipole moment, and n is the e-h density. The component w represents the fractional population difference in the two-level system states and T_1 is the e-h spontaneous recombination time. Due to a strong dependence of the refractive index on carrier density in semiconductor media, a transient spatial population grating results in a corresponding refractive index grating. We have included this effect in the model by an introduction of a parameter which is proportional to the linewidth enhancement factor α of the semiconductor structure [24, 25]. This factor is defined as

$$\alpha = -\frac{\partial Re\chi^{(1)}(\omega)/\partial n}{\partial Im\chi^{(1)}(\omega)/\partial n} \tag{8}$$

The real and imaginary parts of $\chi^{(1)}$ are related to the refractive index and optical gain, respectively. The physical origin of α lies in the asymmetric nature of electron and hole distributions in semiconductors within the energy bands. The value α varies a lot and depends on many factors, including the material parameters of the semiconductor, frequency shift with respect to the optical gain peak, excitation level, device geometry, etc. By contrast to gaseous and solid-state media where α is negligible, it can be far greater than 10 in bulk GaAs/AlGaAs heterostructures at photon energies close to the bandgap [25]. The SR emission exhibits a strong red shift with respect to lasing. This implies that the parameter α under SR is much greater than that during standard lasing in the same structures.

The full-wave vector Maxwell's equations (1)-(3) coupled to the time evolution equations of the two-level quantum system (Eqs. (5)-(7)) are discretized using the Yee, staggered-grid finite-difference algorithm for Maxwell's equations [26]. The algorithm has proven to be very efficient in solving many problems of optics and electromagnetics. The system is solved numerically in the time domain by the standard leapfrog time-advancing scheme. We used the same algorithm for the numerical solution of the equations as described elsewhere [19-21]. At each time step, the Courant stability criterion is fulfilled. The boundary conditions and field reflections at the chip facets are automatically taken into account by the spatial profile of the dielectric constant $\varepsilon(z)$. The modification of the numerical algorithm can be easily performed since the FDTD calculation scheme of Yee has already been adapted to several nonlinear problems of electromagnetism such as the propagation of solitons [22], the SH generation [27], and the self-focusing of beams [28].

Figures 5-7 illustrate the results of the calculations. Since all variables have a very fine sub-μm spatial structure, their distributions inside the device are plotted for visibility for the last few microns at the chip facet. The carrier density and the fundamental harmonic have a sine/cosine-like spatial distribution along the cavity with a $\lambda/2$ and λ periodicity, respectively. The SH electric field has a more complex spatial structure due to the interaction with the transient population and refraction index gratings. Figure 5 presents a typical snap-shot of the inversion during the evolution of an SR pulse. The region of the negligible values at the centre around 40 microns correspond to the position of the absorber section of the device. The transient carrier grating exists for a few picoseconds dying out as the electric field of the SR pulse vanishing inside the cavity.

Figure 6 illustrates the calculated relative permittivity transient grating (a) and the corresponding distribution of the second harmonic field (b). As mentioned above, observed SH power primarily comes from the vicinity of the chip facet due to the large absorption of the semiconductor. The simulations show that the intensity of SH is very low along almost all the cavity length due to strong absorption. The SH power grows nearly exponentially at the facets of the semiconductor chip

The maximum power and the SHG conversion efficiency is strongly affected by transient population and index gratings existing inside the sample. It is expected that the SH power and the conversion efficiency should increase with increasing the initial e-h density at the very beginning of an SR pulse evolution. The calculations demonstrate that the transient gratings are much more pronounced at greater values of n. Figure 7 shows the dependences of the SH peak intensity versus the fundamental peak power and initial e-h density.

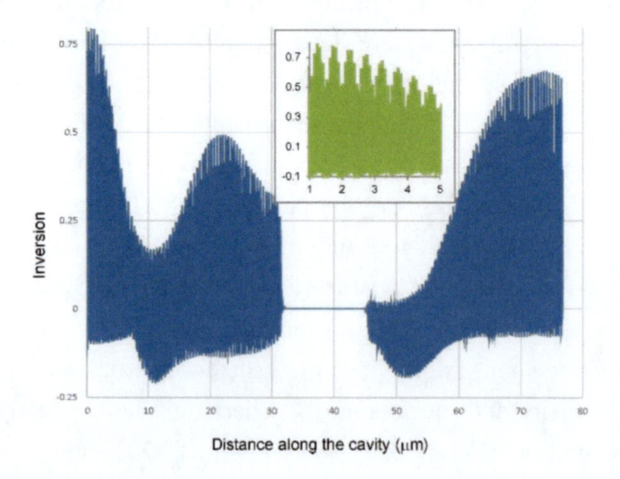

Figure 5. A snap-shot of a typical carrier transient grating. The insert shows the distribution of the inversion at the first 5 microns at the chip facet.

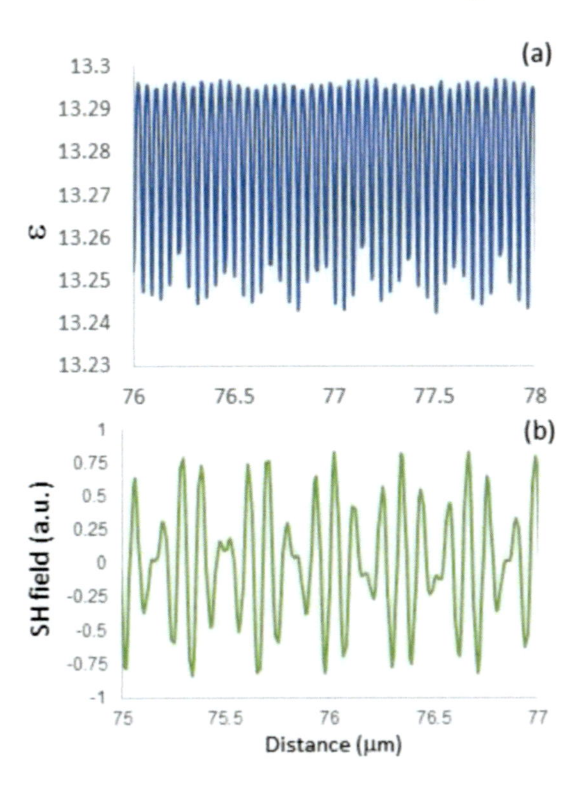

Figure 6. Distribution of the relative permittivity (a) and SH field (b) along the cavity axis at a certain time during SR emission generation.

For relatively small values of P_{ω} the SH power follows the conventional law $P_{2\omega} \propto P_{\omega}^2$. However, at larger P_{ω} the SH intensity increases more rapidly. This is because the peak power of the fundamental harmonic depends strongly on the initial carrier density. At the same time, the strength of both transient population and index gratings increases with carrier density as well. The maximum calculated amplitude of the carrier density variation can be as large as 6 x 10^{18} cm^{-3}, the corresponding refractive index variations being of about 10^{-2} [13]. We attribute the enormous enhancement of the efficiency of the SH generation to the interaction of the electromagnetic field with population and index gratings which are built up in the semiconductor medium during the SR emission generation.

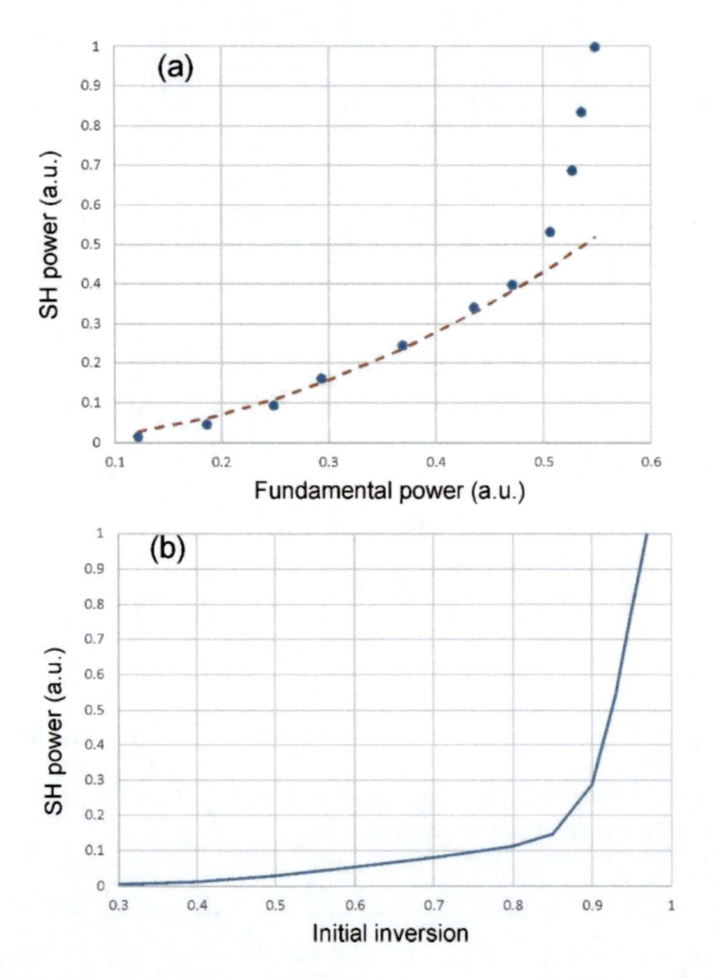

Figure 7. Calculated SH power vs. fundamental power (a) and initial inversion (b). The dashed line in (a) corresponds to the relation $P_{2\omega} \propto P_{\omega}^2$.

DISCUSSION

Internal SH intensity can be easily calculated since SHG has been one of the most studied phenomena in nonlinear optics and the theory of SHG is well-developed. The SH intensity is derived from a wave equation for the electromagnetic field $E_{2\omega}$ [13]. Without depletion of E_{ω} propagating in the z-direction, it reads

$$\frac{\partial E_{2\omega}}{\partial z} = -\frac{i\omega}{\mu_{2\omega} c} d_{eff} E_{\omega}^{2} e^{i\Delta kz} - \gamma E_{2\omega} \tag{9}$$

where d_{eff} is the effective nonlinear coefficient, Δk is the phase mismatch between the fundamental and SH fields, $\mu_{2\omega}$ is the refractive index at the SH wavelength, and c is the velocity of light. The solution of the equation can be readily found and one can write

$$|E_{2\omega}|^{2} \propto \frac{E_{\omega}^{4}}{\Delta k^{2}+\gamma^{2}} \{(cos\Delta kz - e^{-\gamma z})^{2} + (sin\Delta kz)^{2}\} \tag{10}$$

Since the absorption coefficient $\gamma >> 10^{5}$ cm^{-1}, the SH intensity saturates at distances less or around 1 μm. For GaAs one has $d_{eff} = 2.0 \times 10^{-10}$ m/V, $\mu_{\omega} = 3.62$, $\mu_{2\omega} = 5.04$, $\gamma = 2.83 \times 10^{7}$ m^{-1} at the SH wavelength of 440 nm. The solution of Eq. (9) with (10) gives for the SH intensity $I_{2\omega} = c\varepsilon_{0}\mu_{2\omega}|E_{2\omega}|^{2}/2$ (in W/m^{2})

$$I_{2\omega} = \eta I_{\omega}^{2} = \frac{8\pi^{2} d_{eff}^{2}}{\mu_{2\omega}\mu_{\omega}^{2} c\lambda_{\omega}^{2}\varepsilon_{0}(\Delta k^{2}+\gamma^{2})} I_{\omega}^{2} \tag{11}$$

This gives the value of the conversion efficiency $\eta = 8 \times 10^{-9}$ W^{-1} in our case of the emission area of around 2.5×10^{-12} m^{2}. As presented in Table 1, the measured SH power and conversion efficiency are 1-2 orders of magnitude greater than the calculated value.

The exact physical reason behind the ultra-strong SHG under SR generation is not completely clear at the moment. However, one of the possible causes of this effect is the existence of periodic modulation of the e-h density and strong transient population and index gratings which associate the process of SR in semiconductor media. By contrast to standard lasing, when the population grating also exists and the typical modulation of the e-h density is a few percents only, the modulation of e-h density approaches 100% when SR pulses are generated. It has been previously found [8] that coherent transient gratings strongly affect the group refractive index of the medium and cause the superluminal propagation of SR pulses.

A refractive index grating can significantly alter the phase-matching condition of the nonlinear wave interaction. It may permit effective SHG even if the phase-matching condition is not satisfied in a uniform medium of the same background refractive indices at the fundamental and SH frequencies. This phenomenon was first investigated many years ago [29]. A periodic grating rephases the nonlinear polarization and the generated electromagnetic waves. Strong periodic $\lambda/2$ modulation of the e-h density and corresponding modulation of the refractive index can result in quasi-phase matching conditions in a similar manner as in periodically poled materials commonly used for frequency doubling. In addition, it has been previously shown that even a weak spatial periodic modulation of the order of the wavelength of the nonlinear susceptibility $\chi^{(2)}$ of a medium leads to a remarkable enhancement of the SH conversion efficiency [30]. It has been demonstrated that periodic variations of the refractive index of around 10^{-3} – 10^{-2} can result in an increase of the SHG efficiency by more than an order of magnitude at appropriate conditions. Moreover, it has been experimentally proved [31] that the SHG conversion efficiency is enhanced in a periodically stratified GaAs/AlGaAs structures as compared to a spatially uniform medium of the same composition.

It is reasonable to suggest that the periodic modulation of the nonlinear susceptibility of the medium during the SR pulse generation, caused by population and index gratings, leads to a significant (experimentally observed) increase in the internal SHG efficiency. A $\lambda/2$ transient grating of the refraction index distorts the natural dispersion of the material in a similar manner to 1D photonic crystals. As a result, the efficiency of the harmonic doubling is strongly enhanced. A similar effect has been reported by de Ceglia et al. [32] who demonstrated a drastically enhanced SHG in a 1D GaAs subwavelength resonant grating under an external optical pumping. The coupling of the pump field to guided-mode resonances produced by the grating introduces sharp, Fano-like resonances where strong field localization and enhancement take place, leading to the prediction of SHG conversion efficiencies that are 4–5 orders of magnitude larger than in bulk structures.

A completely different effect concerned with abnormal SHG conversion efficiencies in a nonlinear medium was proposed and studied by M.Scalora and co-workers [33]. They considered a similar situation when the absorption of SH in a sample was extremely high. It has been predicted and experimentally observed the enhancement by three orders of magnitude of phase mismatched second and third harmonic generation in a GaAs cavity at wavelengths of 650 and 433 nm, respectively, well above the absorption edge. The authors attributed the enhanced SH conversion efficiencies to a phase-locking mechanism that causes the fundamental harmonic to trap the second and third harmonics and to impress on them its dispersive properties. However, we believe that in our case the periodic modulation of the nonlinear susceptibility of the medium during the SR pulse generation is the dominant mechanism of the enhanced internal SHG conversion efficiency.

CONCLUSION

We have presented experimental and theoretical studies of the generation of ultra-bright internal second harmonic during SR emission generation in semiconductor laser structures. Typical experimentally measured conversion efficiencies proved to be by 1-2 orders of magnitude greater than those previously reported in semiconductor lasers based on bulk and quantum-well materials. For the explanation of the effect, a theoretical model based on nonlinear one-dimensional Maxwell curl equations without taking into consideration the slowly-varying envelope approximation has been developed. The model takes explicitly into account a deep $\lambda/2$ modulation of the carrier density. Due to a strong dependence of refractive index and nonlinear susceptibility of the semiconductor medium on the carrier density corresponding transient gratings are built up. Both e-h density grating and refractive index grating significantly alter the phase-matching condition of the nonlinear wave interaction. The results of theoretical calculations describe well the available experimental data. The formation of transient coherent population and refractive index gratings can explain many features of SR not only the unusually strong internal SHG but superluminal pulse propagation as well.

REFERENCES

[1] Dicke, Robert H. (1954). "Coherence in spontaneous radiation processes." *Physical Review*, *93*, 99–110.

[2] Vasil'ev, Peter P., Penty, Richard V. & White, Ian H. (2013). "Superradiant emission in semiconductor diode laser structures." *IEEE Journal of Selected Topics in Quantum Electronics*, *19*, 1500210.

[3] Cong, Kankan., Zhang, Qi., Wang, Yongrui., Noe, Timothy G., Belyanin, Alexey. & Kono, Junichiro. (2016). "Dicke superradiance in solids." *Journal of the Optical Society of America B*, *33*, C80-C101.

[4] Vasil'ev, Peter P. (2009). "Femtosecond superradiant emission in inorganic semiconductors." *Reports on Progress in Physics*, *72*, 076501.

[5] Xia, Mo., Penty, Richard V., White, Ian H. & Vasil'ev, Peter P. (2012). "Femtosecond superradiant emission in AlGaInAs quantum-well semiconductor laser structures." *Optics Express*, *20*, 8755-8760.

[6] Boiko, Dmitri L. & Vasil'ev, Peter P. (2012). "Superradiance dynamics in semiconductor laser diode structures." *Optics Express*, *20*, 9501-9515.

[7] Olle, Vojtech F., Vasil'ev, Peter P., Wonfor, Adrian., Penty, Richard V. & White, Ian H. (2012). "Ultrashort superradiant pulse generation from a GaN/InGaN heterostructure." *Optics Express*, *20*, 7035-7039.

[8] Vasil'ev, Peter P., Penty, Richard V. & White, Ian H. (2016). "Pulse generation with ultra-superluminal pulse propagation in semiconductor heterostructures by superradiant-phase transition enhanced by transient coherent population gratings." *Light - Science & Applications*, *5*, e16086.

[9] Vasil'ev, Peter P., Olle, Vojtech F., Penty, Richard V. & White, Ian H. (2013). "Long-range order in a high-density electron-hole system at room temperature during superradiant phase transition." *Europhysics Letters*, *104*, 40003.

[10] Malmstrom, M. D., Schlikman, J. J. & Kingstron, R. H. (1964). "Internal second-harmonic generation in gallium arsenide lasers." *Journal of Applied Physics*, *35*, 248-249.

[11] Ogasawara, Nagaatsu., Ito, Ryoichi., Rokukawa, Hiroyuki. & Katsurashima, Wataru. (1987). "Second harmonic generation in an AlGaAs double-heterostructure laser." *Japanese Journal of Applied Physics*, *26*, 1386-1387.

[12] Kanamori, Hideto., Takashima, Shigehiro. & Sakurai, Katsumi. (1991). "Near-infrared diode laser spectrometer with frequency calibration using internal second harmonic." *Applied Optics*, *30*, 3795-3798.

[13] Vasil'ev, Peter P., Penty, Richard V. & White, Ian H. (2018). "Nonlinear optical effects during femtosecond superradiant emission generation in semiconductor laser structures." *Optics Express*, *26*, 26156-26166.

[14] Gross, Michel. & Haroche, Serge. (1982). "Superradiance: an essay on the theory of collective spontaneous emission." *Physics Reports*, *93*(5), 301–396.

[15] MacGillivray, John C. & Feld, Michael S. (1976). "Theory of superradiance in an extended, optically thick medium." *Physical Review A*, *14*, 1169-1189.

[16] Maki, Jeffery J., Malcuit, Michelle S., Raymer, Michael G., Boyd, Robert W. & Drummond, Peter D. (1989). "Influence of collisional dephasing processes on superfluorescence." *Physical Review A*, *40*, 5135-5142.

[17] Vasil'ev, Peter P. (1999). "Role of a high gain of the medium in superradiance generation and in observation of coherent effects in semiconductor lasers." *Quantum Electronics*, *29*, 842-848.

[18] Vasil'ev, Peter P. & Smetanin, Igor V. (2006). "Condensation of electron-hole pairs in a degenerate semiconductor at room temperature." *Physical Review B*, *74*, 125206.

[19] Slavcheva, Gabriela M., Arnold, John M. & Ziolkowski, Richard W. (2004). "FDTD simulation of the nonlinear gain dynamics in active optical waveguides and semiconductor microcavities." *IEEE Journal of Selected Topics in Quantum Electronics*, *10*, 1052-1062.

[20] Nagra, Amit S. & York, Robert A. (1998). "FDTD analysis of wave propagation in nonlinear absorbing and gain media." *IEEE Transactions on Antennas and Propagation*, *46*, 334-340.

[21] Gruetzmacher, Julie A. & Scherer, Norbert F. (2003). "Finite-difference time-domain simulation of ultrashort pulse propagation incorporating quantum-mechanical response functions." *Optics Letters*, *28*, 573-575.

[22] Joseph, Rose M. & Taflove, Allen. A. (1997). "FDTD Maxwell's equations models for nonlinear electrodynamics and optics." *IEEE Transactions on Antennas and Propagation*, *45*, 364-374.

[23] Vasil'ev, Peter P. (2004). "Conditions and possible mechanism of condensation of e-h pairs in bulk GaAs at room temperature." *Physica Status Solidi (b)*, *241*, 1251-1260.

[24] Vasil'ev, Peter. (1995). *Ultrafast diode lasers: fundamentals and applications*. Norwood: Artech House.

[25] Vahala, Kerry., Chiu, L. C., Margalit, Shlomo. & Yariv, Amnon. (1983). "On the linewidth enhancement factor in semiconductor injection lasers." *Applied Physics Letters*, *42*, 631-633.

[26] Yee, Kane S. (1966). "Numerical solution of initial boundary value problems involving Maxwell's equations in isotropic media." *IEEE Transactions on Antennas and Propagation*, *14*, 302–307.

[27] Bourgeade, Antoine. & Freysz, Eric. (2000). "Computational modelling of second-harmonic generation by solution of full-wave vector Maxwell equations." *Journal of the Optical Society of America B*, *17*, 226-234.

[28] Ziolkowski, Richard W. & Judkins, Justin B. (1993). "Full-wave vector Maxwell equation modelling of the self-focusing of ultrashort

optical pulses in a nonlinear Kerr medium exhibiting a finite response time." *Journal of the Optical Society of America B*, *10*, 186–198.

[29] Bloembergen, Nicolaas. & Sievers, Albert J. (1970). "Nonlinear optical properties of periodic laminar structures." *Applied Physics Letters*, *17*, 483–485.

[30] Haus, Joseph W., Viswanathan, Rajesh., Scalora, Michael., Kalocsai, Andre G., Cole, Julian D. & Theimer, James. (1998). "Enhanced second-harmonic generation in a media with a weak periodicity." *Physical Review A*, *57*, 2120-2128.

[31] van der Ziel, Jan P. & Ilegems, Marc. (1976). "Optical second harmonic generation in periodic multilayer GaAs-$Al_{0.3}$ $Ga_{0.7}As$ structures." *Applied Physics Letters*, *28*, 437-439.

[32] de Ceglia, Domenico., D'Aguanno, Mattiucci., Nadia, Vincenti., Maria, A. & Scalora, Michael. (2011). "Enhanced second-harmonic generation from resonant GaAs gratings." *Optics Letters*, *36*, 704-706.

[33] Roppo, Vito., Cojocaru, Crina., Raineri, Fabrice., D'Aguanno, Giuseppe., Trull, Jose., Halioua, Yacine., Raj, Rishi., Sagnes, Isabelle., Vilaseca, Ramon. & Scalora, Michael. (2009). "Field localization and enhancement of phase-locked second- and third-order harmonic generation in absorbing semiconductor cavities." *Physical Review A*, *80*, 043834.

In: Horizons in World Physics. Volume 302 ISBN: 978-1-53616-472-5
Editor: Albert Reimer

Chapter 4

New Fundamentals for Particle Physics: Emergent Spin and Duality of Time

Ivanhoe B. Pestov*
Bogoliubov Laboratory of Theoretical Physics,
Joint Institute for Nuclear Research, Moscow Region, Russia

Abstract

In quantum mechanics and particle physics, Spin is considered as an intrinsic form of the quantum angular momentum of a point particle. The subject of this chapter is to demonstrate that in accordance with the creative original idea of Kronig, Uhlenbek and Goudsmit, we can associate Spin with an intrinsic form of two angular momenta of the quantum Spherical Top. It is shown that the internal symmetry of the Intrinsic Top really exists and it manifests itself as an emergent property of the system of fundamental and simplest geometrical quantities. That is why this phenomenon is called Emergent Spin and to a particle with Emergent Spin one half we put into correspondence a spin field as a carrier space of internal symmetry of the Intrinsic Top. Three families: the electron and the electron neutrino, the muon and the muon neutrino, the tauon and the tauon neutrino can be considered as the evident experimental confirmation of the concept of Emergent Spin since from this new point of view

*Corresponding Author's E-mail: pestov@theor.jinr.ru

they represent different states of a particle with Emergent Spin described by the spin field. The total number of these states equals four and, hence, one more state is predicted. The concept of Spin as an emergent property should be interesting in terms of discussion of possible ways to look for physics beyond the Standard Model since there is no doubt that new physics really exists and we need a clear guidance to the best place to look. To this end, we develop here new fundamentals for the particle physics on the ground of the concept of Emergent Spin and duality of natural Time predicted earlier.

Keywords: space and time, duality of time, lepton-quark symmetry, Emergent Spin

INTRODUCTION

The Standard Model provides an excellent description of what goes on in the physics of elementary particles. However, the trouble begins when we ask the question of why the Standard Model has the features that it does. For example, it holds that there are three different types of leptons: the electron, muon and tau. Why three and not one or four? The Standard Model does not say. Hence, we need to explore a deeper level of nature to discover the answer and to have a clear guidance to the best place to look for physics beyond the Standard Model. In this chapter our goal is to exhibit some physical evidences of the existence of this deeper level, and to this end, we start from the idea of Spin in the context of the representations of classical mechanics with respect to rotational motion (the concept of a "rotating rigid body" also known as the Top).

When Goudsmit and Uhlenbeck proposed the spin hypothesis [1], they had in mind a mechanical picture of Top. This picture had earlier been considered by Kronig as well. However, it was soon recognized that such an approach could not be realized in the sphere of our immediate experience associated with the movement in space. What does it mean? From classical mechanics it follows that for solution of a certain range of problems it is possible not to consider the internal structure of the object in question and to accept that it is at the same internal state (the concept of a "point particle"). With the discovery of quantum mechanics we know that this internal structure exists in the form of a wave field and at this fundamental (field-theoretical) level the internal symmetry plays the same fundamental role as the external symmetry (a transformation of internal symmetry affects the functions of field and does not touch upon the coordinates). Hence, the constructive idea of Kronig, Goudsmith and Uhlenberck can be considered from the new point of view as internal symmetry, which can be put in correspondence to the quantum Spherical Top. The symmetry group of the quantum Spherical Top is well known but, at the fundamental (field - theoretical) level, the idea of Spin as an intrinsic form of two quantum Spherical Top angular momenta was not realized during the period of creation of quantum mechanics. And now, in quantum mechanics and particle physics, Spin is considered as an intrinsic form of quantum angular momentum (one additional degree of freedom instead of degrees of freedom of the Intrinsic Top which will be described in detail in what follows). One can associate this approach with a visual picture of rotation, when the axis of rotation is constant during the motion. In general, it is not the case. The rotations with respect to a fixed axis are radically different from a rotation with respect to a stationary point, where the axis of rotation is not fastened in space (the notion of instantaneous axis of rotation). It should be noted that the idea of Spin as an intrinsic being of quantum angular momentum of a point particle was realized in the Dirac theory.

Now our goal is to recognize a deeply hidden structure that can be put in correspondence to the intrinsic form of two angular momenta of the quantum Spherical Top. It is shown that the idea of Spin as an Intrinsic Top is realized on a set of simplest geometrical quantities, which themselves do not exhibit this property. Hence, Spin as an intrinsic being of two angular momenta of the quantum Spherical Top is an emergent property and this is the reason to call this phenomenon Emergent Spin. A field which we put in correspondence to a particle with Emergent Spin one half will be called spin field. The spin field is a set of fields, the meaning of which is not determined by the meaning of the fields it consists of. The theory of the spin field which has internal degrees of freedom of the Intrinsic Top is developed in detail and can be exhibited as a new form of the Standard Model. The local internal transformations (linear operators), which allow one to consider the spin field as a system and Spin as its emergent property, form a group of general linear transformations $GL(2^n, R.)$ We are striving to give all definitions at any n, but of course the case $n = 4$ is preferred. It is shown that Emergent Spin can be visible as an Intrinsic Top.

The Intrinsic Top is defined by two dual sets of commuting operators which form the natural basis in the space of linear operators in question and define two dual subgroups S and $\widetilde{S}$ of $GL(2^n, R)$ as well. The transformations of these subgroups commute with each other and the spin field is the space of their two-valued representation. The spin field describes the physical entity retaining the elementarity and is considered to be characterized by its equation and the electrical charge in one case and the pseudo-charge, electric charge and neutrino charge in the other case. The neutrino charge is a new quantum number which is tightly connected with the concept of orientation of space and naturally explains the the so called pairing of the electron and neutrino in the Standard Model.

It is shown that from the concept of natural Time, first clearly defined in the work [2], there follows the duality of Time. The first Time provides deeper fundamentals for the special theory of relativity. The dual theory describes the general rotational motion as a motion in the dual Time, where any particle is characterized by two angular momenta of the Spherical Top and angular momenta of the Intrinsic Top. This phenomenon is called Fundamental Spin. We demonstrate that in a certain sense the well-known idea of a "rotating rigid body" (also mentioned as Top) of classical mechanics is as fundamental as the idea of a "mass point", i.e., the first concept can be reduced to the second one at the fundamental (field -theoretical) level in the framework of the geometry in which point is defined as a point on the three-dimensional sphere. The dual Time plays a key role in this consideration where everything rotates (the phenomenon of Fundamental Spin). Thus, from the duality of Time it follows that any known particle can be put in correspondence to a dual particle moving in the dual Time (dual particles live in the space of constant positive curvature). Hence, it is natural to put forward the idea of dual approach to the world of elementary particles which can explain the existence of leptons and quarks, lepton-quark symmetry and confinement (if we identify dual particles with quarks). The observed generations of quarks and leptons can be considered in this case as different states of the particle that we put in correspondence to the complex spin field, and the number of these generations equals four. Thus, the concept of Emergent Spin as appropriate realization of the fundamental idea of Kronig, Uhlenbek and Goudsmit can be interesting in terms of discussion of possible ways to look for physics beyond the Standard Model.

CONCEPTS OF SPACE AND TIME

In this section, we exhibit a very general and deep property of natural Time introduced in the work [2]. We start from the definition of natural geometry and natural Time to demonstrate the duality of Time.

Since the field of real numbers R is continuous and irrelevant to all forms of physical matter, we can define on this ground continuous and irrelevant natural geometry R^n, in which a point is defined as an n-tuple of real numbers

$$x = (x^1, x^2, \cdots x^n),$$

and the distance function is introduced as usual

$$d(x, y) = \sqrt{(x^1 - y^1)^2 + (x^2 - y^2)^2 + \cdots + (x^n - y^n)^2}.$$

This function has a deep meaning since the field of real numbers R is invariant with respect to the transformations $x \Rightarrow x + a$ and from this point of view x is not observable but $x - y$ is the case. In accordance with this observation we give the definition of a vector of the reference space R^n. Let P and Q be the points

$$x = (x^1, x^2, \cdots x^n)$$

and

$$y = (y^1, y^2, \cdots y^n),$$

then the vector $\mathbf{PQ}$ has the components

$$\mathbf{PQ} = (x^1 - y^1, x^2 - y^2, \cdots x^n - y^n)$$

and the distance function is considered as its length.

It is clear that R^1, R^2, R^3 provide a new irrelevant representation of such things as the Euclidian straight line, plane and space, respectively. However, R^1, R^2, R^3 admit simple and clear generalization and, hence, R^n is a very important geometry which can be considered as the underlying structure of any investigation in the field of geometry and physics. On the basis of natural geometry more complicated geometries may be constructed in which a point is defined as a point of some n-dimensional surface in the space R^N, $n < N$. These generalized geometries can be put in correspondence to the states of a full system of fields (a system which includes the gravitational field) [2].

It is evident that the variables (Cartesian coordinates) $x^1, x^2, \cdots x^n$ in the definition of point R^n should be considered on an absolutely equal footing. Hence, it is unclear how to introduce the so-called space coordinates and time coordinate (a space-time structure of R^n) in the framework of natural geometry alone. To do this, we need to give a definition of natural Time as an entity which is tightly connected with all natural dynamical processes and is as simple as possible from a geometrical point of view. To make it easier to perceive the definition given in [2] and below, let us appeal to physical intuition. We know very well the physical phenomena connected with the temperature and pressure difference. We speak about the gradient of temperature and pressure and presuppose that values of these physical quantities are known for any point of some region of the Euclidian space. From a geometrical point of view we deal with a scalar field that is invariant with respect to all admissible transformations of coordinates. Now it is natural to suppose that there is a field of moments of Time and an area of phenomena defined by the gradient of Time.

Definition: a moment of natural Time is a number that we put in correspondence to any point of the reference space R^n. Hence, a moment of Time is defined by the equation

$$t = f(x^1, x^2, \cdots x^n) = f(x).$$

In order to understand the nature of Time, it is very important to clarify that a moment of Time is invariant with respect to general coordinate transformations

$$\bar{x}^i = \bar{x}^i(x^1, x^2, \cdots, x^n), \quad x^i = x^i(\bar{x}^1, \bar{x}^2, \cdots, \bar{x}^n), \quad i = 1, 2, \cdots n,$$

where $x^1, x^2, \cdots, x^n$ are the Cartesian coordinates and $\bar{x}^1, \bar{x}^2, \cdots, \bar{x}^n$ are the new ones.

All points of the reference space that correspond to the same moment of Time t constitute physical space $S(t)$. A point of $S(t)$ is defined by the equation

$$f(x^1, x^2, \cdots x^n) = f(x) = t = constant.$$

This one-parameter family of physical spaces defines causality or determinism of physical reality itself.

The gradient of Time is the vector field $\mathbf{t}$ with the components

$$t^i = (\nabla f)^i = g^{ij}\partial_j f = g^{ij}t_j,$$

where g^{ij} are the contravariant components of the Riemann positive definite metric field $g_{ij}(x)$, which we have put in correspondence to the gravitational field [2].

The gradient of Time defines fundamental discrete internal symmetry known as bilateral symmetry. A pair of vector fields $\mathbf{v}$ and $\overline{\mathbf{v}}$ has bilateral symmetry if the sum of these fields is collinear to the gradient of Time and their difference is orthogonal to it,

$$\overline{\mathbf{v}} + \mathbf{v} = \lambda \mathbf{t}, \quad (\overline{\mathbf{v}}|\mathbf{t}) = (\mathbf{v}|\mathbf{t}),$$

where

$$(\mathbf{v}|\mathbf{w}) = g_{ij} v^i w^j = v^i w_i$$

is a scalar product. It is very important that the bilateral symmetry be represented as a linear transformation (reflection)

$$\overline{v}^i = R^i_j v^j, \quad R^i_j = 2t^i t_j - \delta^i_j.$$

The gradient of Time and the metric field are invariant under reflection since

$$R^i_j t^j = t^i, \quad R^i_k R^j_l g_{ij} = g_{kl}.$$

The bilateral symmetry defines the causal structure of the physical world since the auxiliary metric

$$\overline{g}_{ij} = 2t_i t_j - g_{ij} = g_{ik} R^k_j, \quad \overline{g}^{ij} = 2t^i t^j - g^{ij}$$

provides an effective method of consideration of the dynamical processes through the introduction of natural Time into the Lagrangians (and the equations) of the fundamental physical fields [2]. We present an idea that bilateral symmetry is strict and fundamental symmetry of nature and, hence, in all physical processes one cannot distinguish the right-hand sided physical quantity from the left-hand sided one.

From the consideration of the bilateral symmetry it follows that in the geometrical (coordinate independent) form the time reversal invariance means that the theory is invariant with respect to the transformations

$$T: \quad t^i \to -t^i.$$

It is clear that the transformation T has meaning if and only if the domains of values of the potentials $f(x)$ and $-f(x)$ coincide. It is clear that the theory will be time reversal invariant if the gradient of the temporal field appears in all formulae only as an even number of times, like $t^i t^j$.

Since the scalar temporal field enters into the Lagrangians of the physical fields in the form of the gradient of the scalar field

$$t_i = \partial_i f(u),$$

the laws of the unified physics are invariant with respect to transformations of the form

$$f(x) \Rightarrow f(x) + a,$$

where a is a constant. This symmetry defines the law of energy conservation as a fundamental physical law of the universe which is true in all cases.

The potential $f(x)$ of natural Time is a solution to the equation

$$D_{\mathrm{t}} f = (\nabla f)^2 = g^{ij} \frac{\partial f}{\partial x^j} \frac{\partial f}{\partial x^j} = 1, \tag{1}$$

since this equation can be considered as the definition of uniformity of natural Time. Other mathematical arguments in favour of this equation are also impressible. Let $dx^i = t^i dt$, $f(x + dx) = t + dt$, $f(x) = t$, then $df(x) = t_i t^i dt = dt$ and, hence, $t_i t^i = 1$. This is equation (1). Further, we consider the differential operator $D_{\mathrm{t}} = t^i \partial_i$ defined by the gradient of natural Time t^i and its exponent $\exp(aD_{\mathrm{t}}) = 1 + aD_{\mathrm{t}} + \frac{a^2}{2}(D_{\mathrm{t}})^2 + \cdots$. We put forward the natural demand that transformation $f(x) \Rightarrow f(x) + a$ is generated by the exponent of the gradient of natural Time t^i, and from the equation $\exp(aD_{\mathrm{t}})f(x) = f(x) + a$ we again derive equation (1). Taking into account the possibility of changing the scale, we also subordinate the potential $f(x)$ of natural Time to the equation

$$f(\lambda x^1, \lambda x^2, \cdots \lambda x^n) = \lambda f(x^1, x^2, \cdots x^n). \tag{2}$$

The fundamental observation (from a physical point of view) reads that equation (1) has not only general solution but also a special solution known as the function of the geodesic distance [3]. This means that there are two different Times in nature and, hence, two different kinds of natural dynamical processes. Below we consider a strict and fundamental realization of this statement.

In what follows we consider the four-dimensional reference space R^4 with the metric $ds^2 = g_{ij} dx^i dx^j = (dx^1)^2 + (dx^2)^2 + (dx^3)^2 + (dx^4)^2, \quad g_{ij} = \delta_{ij}$.

We consider equations (1) and (2) under these conditions and, hence, look for solutions to the equations

$$\left(\frac{\partial f}{\partial x^1}\right)^2 + \left(\frac{\partial f}{\partial x^2}\right)^2 + \left(\frac{\partial f}{\partial x^3}\right)^2 + \left(\frac{\partial f}{\partial x^4}\right)^2 = 1,$$

$$f(\lambda x^1, \lambda x^2, \lambda x^3, \lambda x^4) = \lambda f(x^1, x^2, x^3, x^4).$$

In accordance with our general statement, these equations have two solutions:

$$f(x) = a_i x^i = a_1 x^1 + a_2 x^2 + a_3 x^3 + a_4 x^4,$$

where $\mathbf{a} = (a^1,\ a^2,\ a^3,\ a^4)$ is a unit constant vector $(\mathbf{a}|\mathbf{a}) = 1$, and

$$f(x) = \sqrt{(x^1)^2 + (x^2)^2 + (x^3)^2 + (x^4)^2}.$$

From the equations

$$f(x) = a_i x^i = a_1 x^1 + a_2 x^2 + a_3 x^3 + a_4 x^4 = t = constant,$$

and

$$f(x) = \sqrt{(x^1)^2 + (x^2)^2 + (x^3)^2 + (x^4)^2} = \tau = constant$$

we see that in one case physical space is the familiar three-dimensional Euclidian space E^3 and in the other case the new physical space is the three-dimensional Riemannian space of constant positive curvature, i.e., the 3d-sphere S^3. The physical (mass) points are to be identified with the points belonging to the three-dimensional Euclidian space E^3, but the points belonging to the 3d-sphere S^3 should be put in correspondence to the Spherical Top. Indeed, the symmetries of the Euclidian space can be composed of translations and rotations and the symmetries of the 3d-sphere S^3 coincide with those of the Spherical Top which will be considered below. In other words, geometrical points in the Euclidian and Riemannian spaces have a different physical meaning. Thus, from the duality of Time it follows that any known particle can be put in correspondence to a dual particle moving in the dual Time. We see that there are two space-time structures (two different Times) on the same reference space R^4. It will be shown that the first space-time structure of R^4 uncover the essence of the special theory of relativity, and the dual space-time structure describes natural rotation as a motion in the dual Time. Hence, it is natural to put forward the idea of dual approach to the world of elementary particles which can explain the existence of leptons and quarks, lepton-quark symmetry and confinement (if we identify dual particles with quarks).

FUNDAMENTAL PHYSICAL FIELDS

The surfaces of any possible dimensions form the geometrical structure of the reference space R^n. Following the fundamental ideas of Riemann and Einstein, we consider an important class of fundamental physical fields which are defined by the functionals on the set of p-dimensional surfaces in the reference space.

The $p-$ dimensional surfaces in the parametric form are defined by the equations

$$S_p : x^i = x^i(u^1, u^2, \cdots u^p) \quad (p = 0, 1, 2, 3, ...n).$$

The Case $p = 0$ represents a surface of dimension zero or a point P and "integral" from a scalar field $a(x)$ with respect to this surface equal to $a(P)$. Putting $p = 1$, we find a curve $\gamma : x^i = x^i(u)$. Thus, we consider the functionals

$$F_p = \int a_{i_1 \cdots i_p}(x) \frac{\partial x^{i_1}}{\partial u^1} \cdots \frac{\partial x^{i_p}}{\partial u^p} du^1 \cdots du^p,$$

to introduce into consideration alongside with the positive-definite metric field $g_{ij}(x)$ a scalar field $a(x)$, a covector field $a_i(x)$ and antisymmetrical tensor fields of different rank $a_{ij}(x)$, $a_{ijk}(x)$, $a_{ijk \cdots l}(x)$ as elements of a unique self-consistent structure. The Stoke theorem should be considered in this context as an additional convincing argument.

Remember that Riemann introduced a new field (the positive-definite metric field $g_{ij}(x)$) as a functional

$$F(\gamma, g) = \int_{u_0}^{u_1} \sqrt{g_{ij} \frac{dx^i}{du} \frac{dx^j}{du}} du.$$

We stress here the fundamental role of this field. The symmetry principles made their appearance in the twentieth century physics with identification of the invariance group of the Maxwell equations. With this as a precedent, symmetries took on a character in minds of physicists as a priori principles of universal validity. The natural generalization appears as a fundamental principle that basic laws for the simplest forms of matter should be defined by the widest possible groups of transformations. The principle of reparametrization invariance (general relativity of Einstein) represents one of these fundamental groups. And now we state that to be reparametrization invariant, any theory of fields must include the metric field $g_{ij}(x)$. But this principle proved insufficient to reach the goal at which field physics is aimed: a unified field theory deriving all laws from one common, harmonic and self-consistent structure of the world. Some elements of this structure were mentioned immediately above, and to complete this list, let us define one more fundamental physical field.

To any curve γ we can put in correspondence a set of vectors parallel along this curve. This set of parallel vectors is defined by the system of equations

$$\frac{dv^i}{dt} + P^i_{jk} v^k \frac{dx^j}{dt} = 0$$

and the fundamental physical field with the components $P^i_{jk}(x)$ known as linear connection.

Thus, the set of fundamental physical fields, tightly connected with the geometrical structure of the reference space, is very restricted. Let us enumerate these basis fields:

1) The positive-definite metric field $g_{ij}(x)$;
2) the field of parallel displacement $P^i_{jk}(x)$;
3) the scalar and covariant vector fields, antisymmetric covariant tensor fields. For our goals it is useful to show the last set of fields as a 2^n- tuple

$$\Big(a(x),\ a_i(x),\ a_{ij}(x), \cdots a_{ijk\cdots l}(x)\Big).$$

The other notable argument in favor of the set of fundamental physical fields outlined above gives a concept of potential field which was introduced in the most general form in [4]. If we take the components of the symmetrical covariant tensor field g_{ij} and form its derivatives ($\partial_i g_{jk}$), then these derivatives are neither the components of the tensor nor of any geometrical object. However, from g_{ij} and these partial derivatives one can form (with the help of algebraic operations only) a new geometrical object

$$\Gamma^i_{jk} = \frac{1}{2} g^{il} (\partial_j g_{kl} + \partial_j g_{kl} - \partial_l g_{jk}),$$

which is called the Christoffel connection where g^{il} are contravariant components of g_{ij}. Now we can formalize this particular case and give general definition of the potential field.

If some geometrical object (or a geometrical quantity) is given and from the components of this object and its partial derivatives one can form (using the algebraic operations only) a new geometrical object (or geometrical quantity), then we deal with a new geometrical quantity that will be called a potential field. The potential field is characterized by the potential P and the strength H and in what follows it will be written in the form (P,H). The connection between the potential and the strength is then called a natural derivative and in a symbolic form can be written as follows $H = \partial P$.

If we go back to our starting point, g_{ij} is a potential and Γ^i_{jk} is a strength of the potential field (g, Γ).

Further, let us consider the Riemann tensor of the connection P^i_{jk}

$$B_{ijl}{}^k = \partial_i P^k_{jl} - \partial_j P^k_{il} + P^k_{im} P^m_{jl} - P^k_{jm} P^m_{il}.$$

Now we go back to the definition of the potential field and see that our field geometrical quantity defines a new potential field (P, B).

The natural derivative of the elements of 2^n- tuple

$$(a, a_i, a_{ij}, \cdots a_{ijk\cdots l})$$

is the external derivative which defines the following set of potential fields

$$(a,\, \partial_i a),\, (a_i,\, \partial_i a_j - \partial_j a_i),\, (a_{ij}, \partial_i a_{jk} + \partial_j a_{ki} + \partial_k a_{ij}), \cdots (a_{ijk\cdots l},\, 0).$$

Finally, we pay attention to that the potential field $(f,\, \partial_i f)$ defines the natural Time.

IDEA OF INTRINSIC TOP

Remember that the constructive idea of Kronig, Goudsmith and Uhlenberck can be considered from a new point of view as internal symmetry, which can be put in correspondence to the quantum Spherical Top. The symmetry group of the quantum Spherical Top is well known and it has a transparent geometrical explanation from a four-dimensional point of view [5] (It was detected in that the Spherical Top is a simple geometrical construction strictly defined by a moving point in the four-dimensional Euclidian space). This group is defined by the following dual laws:

$$[L_j, L_k] = i\varepsilon_{jkl}L_l, \quad [\widetilde{L}_j, \widetilde{L}_k] = i\varepsilon_{jkl}\widetilde{L}_l, \quad [L_j, \widetilde{L}_k] = 0. \tag{3}$$

$$L_1^2 + L_2^2 + L_3^2 = \widetilde{L}_1^2 + \widetilde{L}_2^2 + \widetilde{L}_3^2 \tag{4}$$

which represent the Lie algebras of the two dual groups G and $\widetilde{G}$ (symmetry groups of quantum Spherical Top). The transformations of G and $\widetilde{G}$ commute with each other and in the concept of Intrinsic Top we need to reproduce this structure and its intrinsic form:

$$[S_j, S_k] = i\varepsilon_{jkl}S_l, \quad S_1^2 + S_2^2 + S_3^2 = s(s+1) = \frac{3}{4}, \tag{5}$$

$$[\widetilde{S}_j, \widetilde{S}_k] = i\varepsilon_{jkl}\widetilde{S}_l, \quad \widetilde{S}_1^2 + \widetilde{S}_2^2 + \widetilde{S}_3^2 = \frac{3}{4}, \quad [S_j, \widetilde{S}_k] = 0. \tag{6}$$

At the fundamental (field - theoretical) level the idea of Spin as an intrinsic form of the quantum Spherical Top angular momenta was not realized during the period of foundation for quantum mechanics. And now, in quantum mechanics and particle physics, Spin is considered as an intrinsic form of orbital angular momentum of a point particle with the laws analogous to those of the quantum orbital angular momentum:

$$[S_j, S_k] = i\varepsilon_{jkl}S_l, \quad S_1^2 + S_2^2 + S_3^2 = s(s+1) = \frac{3}{4}.$$

To these laws we put in correspondence a visual picture of rotation, when the axis of rotation is constant during the motion. In general, it is not the case since the rotations with respect to a fixed axis is of a radically different from the rotations with respect to a stationary point, where the axis of rotation is not fastened in space (the notion of instantaneous axis of rotation). It should be noted that the idea of Spin as an intrinsic being of quantum angular momentum of a point particle was realized by Dirac.

Now our goal is to recognize a deeply hidden structure that can be put in correspondence to equations (5) and (6). It will be shown that the idea of Spin as an intrinsic form of the quantum Spherical Top angular momenta can be realized on a set of simplest geometrical quantities defined above, which themselves do not exhibit this property. Hence, Spin as an intrinsic being of the quantum Spherical Top angular momenta is an emergent property and this is the reason to call this phenomenon Emergent Spin. The field which we will put in correspondence to the Emergent Spin will be called spin field. In connection with the concept of Emergent Spin it should be noted that A. M. Baldin did pay attention to the fact that Reductionism was not the all-inclusive principle of nature and he had the constructive ideas in this direction.

INTRINSIC TOP

In the dimension $n = 4$ natural geometry is defined by a set of real numbers R and a point of this geometry is defined as a 4-tuple of real numbers (in the general case, n-tuple)

$$x = (x^1, x^2, x^3, x^4).$$

More complicated geometries may be constructed on the basis of natural geometry. All points of natural geometry form reference space R^4. A set of simplest (irreducible) geometrical quantities on R^4 consists of the scalar field $a(x)$, covariant vector field $a_i(x)$, and the antisymmetric covariant tensor fields $a_{ij}(x)$, $a_{ijk}(x)$ and $a_{ijkl}(x)$, which can be arranged as a tuple:

$$\mathbf{A} = (a, a_i, a_{ij}, a_{ijk}, a_{ijkl}), \quad i, j, k, l = 1, 2, 3, 4. \tag{7}$$

The set of tuples in question can be considered as linear space of dimension 2^4. With the general definition of linear space we have nothing to do with the preferred dimension of this space. However, it will be shown that from geometrical and group-theoretical points of view dimension 2^4 (in general 2^n) is singled out.

Let us exhibit a backbone property that gives an opportunity to interpret 2^4-tuple (7) as a system. We consider linear transformations (operators) $\hat{\mathrm{T}} : \overline{\mathbf{A}} = \hat{\mathrm{T}}\mathbf{A}$ which are defined by a set of totally antisymmetrical tensor fields of type (p,q), $T^{i_1\cdots i_p}_{j_1\cdots j_q}, \quad (p, q = 0, 1, 2, 3, 4)$. We put

$$\overline{\mathbf{A}} = (\overline{a}, \overline{a}_i, \overline{a}_{ij}, \overline{a}_{ijk}, \overline{a}_{ijkl}), \quad (i, j, k, l = 1, 2, 3, 4)$$

and have

$$\overline{a} = Ta + T^m a_m + \frac{1}{2!}T^{mn}a_{mn} + \frac{1}{3!}T^{mnp}a_{mnp} + \frac{1}{4!}T^{mnpq}a_{mnpq},$$

$$\overline{a}_i = T_i a + T_i^m a_m + \frac{1}{2!} T_i^{mn} a_{mn} + \frac{1}{3!} T_i^{mnp} a_{mnp} + \frac{1}{4!} T_i^{mnpq} a_{mnpq},$$

$$\overline{a}_{ij} = T_{ij} a + T_{ij}^m a_m + \frac{1}{2!} T_{ij}^{mn} a_{mn} + \frac{1}{3!} T_{ij}^{mnp} a_{mnp} + \frac{1}{4!} T_{ij}^{mnpq} a_{mnpq},$$

$$\overline{a}_{ijk} = T_{ijk} a + T_{ijk}^m a_m + \frac{1}{2!} T_{ijk}^{mn} a_{mn} + \frac{1}{3!} T_{ijk}^{mnp} a_{mnp} + \frac{1}{4!} T_{ijk}^{mnpq} a_{mnpq},$$

$$\overline{a}_{ijkl} = T_{ijkl} a + T_{ijkl}^m a_m + \frac{1}{2!} T_{ijkl}^{mn} a_{mn} + \frac{1}{3!} T_{ijkl}^{mnp} a_{mnp} + \frac{1}{4!} T_{ijkl}^{mnpq} a_{mnpq}.$$

The identical transformation $\hat{\mathrm{E}}$ is defined by the following natural conditions

$$E^{i_1 \cdots i_p}_{j_1 \cdots j_q} = 0, \quad \text{if} \quad p \neq q, \quad E^{i_1 \cdots i_p}_{j_1 \cdots j_p} = \delta^{i_1 \cdots i_p}_{j_1 \cdots j_p},$$

where $\delta^{i_1 \cdots i_p}_{j_1 \cdots j_p}$ is the generalized Kronecker delta.

The so defined local internal transformations form the group $GL(2^4, \mathbf{R})$ which allows one to consider the 2^4- tuple (7) as a unit. Now it is time to speak about an emergent property. The emergent behavior (or emergent property) can appear when a number of simple entities live in an environment forming their unexpected behavior. The number of simple entities is known (see equation (7)). We also know something about the “environment”. By that I mean the linear space of the operators $\hat{\mathrm{T}}$ (defined above). But we know nothing about the hidden properties of this “environment”. To detect this underlying structure, we need to introduce one very important element: a positive-definite Riemannian metric field $g_{ij}(x)$, which is a key to realize the natural properties of Intrinsic Top and detect the concept of Emergent Spin as well. In this sense, the concept of Emergent Spin has a metric nature.

Indeed, the metric field $g_{ij}(x)$ gives an opportunity to construct a natural general covariant basis in the linear space of the operators $\hat{\mathrm{T}}$ and define the Intrinsic Top as two dual sets of commuting operators. They provide the needed basis and define two dual subgroups S and $\widetilde{S}$. The transformations of S and $\widetilde{S}$ commute with each other and the spin field $\mathbf{A}$ is the space of the two-valued representation of these dual groups. We consider this two-valuedness as direct rational evidence of the existence of Emergent Spin.

To construct the above basis, let us consider the natural algebraic operators $\overline{\mathbf{A}} = \hat{\mathrm{E}}_{\mathbf{v}} \mathbf{A}$ and $\overline{\mathbf{A}} = \hat{\mathrm{I}}_{\mathbf{v}} \mathbf{A}$ defined by the vector field v^i as follows:

$$\hat{\mathrm{E}}_{\mathbf{v}} : \overline{\mathbf{A}} = \hat{\mathrm{E}}_{\mathbf{v}} \mathbf{A} = (0,\ v_i a,\ v_{[i} a_{j]},\ v_{[i} a_{jk]},\ v_{[i} a_{jkl]}),$$

$$\hat{\mathrm{I}}_{\mathbf{v}} : \overline{\mathbf{A}} = \hat{\mathrm{I}}_{\mathbf{v}}\mathbf{A} = (v^m a_m, \, v^m a_{mi}, \, v^m a_{mij}, \, v^m a_{mijk}, \, 0),$$

where the square brackets $[\cdots]$ denote the process of alternation and $v_i = g_{ij}v^j$. For any vector fields v^i and w^i we have

$$\hat{\mathrm{I}}_{\mathbf{v}}\hat{\mathrm{E}}_{\mathbf{w}} + \hat{\mathrm{E}}_{\mathbf{w}}\hat{\mathrm{I}}_{\mathbf{v}} = (\mathbf{v}|\mathbf{w}) \cdot \hat{\mathrm{E}},$$

where $(\mathbf{v}|\mathbf{w}) = g_{ij}v^iw^j$ is a positive-definite scalar product. We mention also the evident relations

$$\hat{\mathrm{E}}_{\mathbf{v}}\hat{\mathrm{E}}_{\mathbf{w}} + \hat{\mathrm{E}}_{\mathbf{w}}\hat{\mathrm{E}}_{\mathbf{v}} = 0, \quad \hat{\mathrm{I}}_{\mathbf{v}}\hat{\mathrm{I}}_{\mathbf{w}} + \hat{\mathrm{I}}_{\mathbf{w}}\hat{\mathrm{I}}_{\mathbf{v}} = 0.$$

To complete, let us introduce the numerical diagonal operator $\hat{\mathrm{Z}}$ that is defined by the conditions

$$Z^{i_1 \cdots i_p}_{j_1 \cdots j_q} = 0, \quad \text{if} \quad p \neq q, \quad Z^{i_1 \cdots i_p}_{j_1 \cdots j_p} = (-1)^p \delta^{i_1 \cdots i_p}_{j_1 \cdots j_p}.$$

From the definition of $\hat{\mathrm{Z}}$, we immediately have the following relations:

$$\hat{\mathrm{E}}_{\mathbf{v}}\hat{\mathrm{Z}} + \hat{\mathrm{Z}}\hat{\mathrm{E}}_{\mathbf{v}} = 0, \quad \hat{\mathrm{I}}_{\mathbf{v}}\hat{\mathrm{Z}} + \hat{\mathrm{Z}}\hat{\mathrm{I}}_{\mathbf{v}} = 0, \quad \hat{\mathrm{Z}}^2 = \hat{\mathrm{E}}. \tag{8}$$

Now we introduce the fundamental operators:

$$Q_{\mathbf{v}} = \hat{\mathrm{E}}_{\mathbf{v}} - \hat{\mathrm{I}}_{\mathbf{v}}, \quad \widetilde{Q}_{\mathbf{v}} = (\hat{\mathrm{E}}_{\mathbf{v}} + \hat{\mathrm{I}}_{\mathbf{v}})\hat{\mathrm{Z}} \tag{9}$$

which define the Intrinsic Top and together with this the Emergent Spin.

From the definition of the operators $Q_{\mathbf{v}}$ and $\widetilde{Q}_{\mathbf{v}}$ it follows that

$$Q_{\mathbf{v}}Q_{\mathbf{w}} + Q_{\mathbf{w}}Q_{\mathbf{v}} = -2(\mathbf{v}|\mathbf{w}) \cdot \hat{\mathrm{E}}, \quad \widetilde{Q}_{\mathbf{v}}\widetilde{Q}_{\mathbf{w}} + \widetilde{Q}_{\mathbf{w}}\widetilde{Q}_{\mathbf{v}} = -2(\mathbf{v}|\mathbf{w}) \cdot \hat{\mathrm{E}},$$

and, hence, $Q^2_{\mathbf{v}} = \widetilde{Q}^2_{\mathbf{v}} = -(\mathbf{v}|\mathbf{v}) \cdot \hat{\mathrm{E}}$. Besides, we also deduce one very important relation

$$\widetilde{Q}_{\mathbf{v}}Q_{\mathbf{w}} = Q_{\mathbf{w}}\widetilde{Q}_{\mathbf{v}},$$

that is fulfilled at any vector fields $\mathbf{v}$ and $\mathbf{w}$ and defines the Intrinsic Top as well.

Indeed, let us introduce the dual operators

$$Q_{\mathbf{v}\wedge\mathbf{w}} = \frac{1}{2!}(Q_{\mathbf{v}}Q_{\mathbf{w}} - Q_{\mathbf{w}}Q_{\mathbf{v}}), \quad \text{and} \quad \widetilde{Q}_{\mathbf{v}\wedge\mathbf{w}} = \frac{1}{2!}(\widetilde{Q}_{\mathbf{v}}\widetilde{Q}_{\mathbf{w}} - \widetilde{Q}_{\mathbf{w}}\widetilde{Q}_{\mathbf{v}}).$$

The above formula can be generalized as follows. We take four linear independent vector fields $\mathbf{u}, \mathbf{v}, \mathbf{w}, \mathbf{z}$ and introduce the operators which are defined as an alternated product of the operators $Q_{\mathbf{u}}, Q_{\mathbf{v}}, Q_{\mathbf{w}}, Q_{\mathbf{z}}$ and $\widetilde{Q}_{\mathbf{u}}, \widetilde{Q}_{\mathbf{v}}, \widetilde{Q}_{\mathbf{w}}, \widetilde{Q}_{\mathbf{z}}$. We see that the operators $Q_{\mathbf{u}\wedge\mathbf{v}\cdots\wedge\mathbf{z}}$ and $\widetilde{Q}_{\mathbf{u}\wedge\mathbf{v}\cdots\wedge\mathbf{z}}$ commute with each other and all their possible products form the needed general covariant basis in the space of linear operators $\hat{\mathrm{T}}$. The total number of these operators is equal to $2^4 \cdot 2^4$ (including the identical operator $\hat{\mathrm{E}}$). Besides, the dual operators

$$Q_{\mathbf{u}},\ Q_{\mathbf{v}},\ Q_{\mathbf{w}},\ Q_{\mathbf{z}},$$

$$\widetilde{Q}_{\mathbf{u}},\ \widetilde{Q}_{\mathbf{v}},\ \widetilde{Q}_{\mathbf{w}},\ \widetilde{Q}_{\mathbf{z}}$$

and their alternative products define the dual groups S and $\widetilde{S}$ mentioned above.

At last, let us consider three orthogonal and unit vector fields $\mathbf{u}$, $\mathbf{v}$ and $\mathbf{w}$, $(\mathbf{u}|\mathbf{v}) = (\mathbf{u}|\mathbf{w}) = (\mathbf{v}|\mathbf{w}) = 0$, $(\mathbf{u}|\mathbf{u}) = (\mathbf{v}|\mathbf{v}) = (\mathbf{w}|\mathbf{w}) = 1$ and put

$$S_1 = \frac{i}{2} Q_{\mathbf{v}\wedge\mathbf{w}}, \quad S_2 = \frac{i}{2} Q_{\mathbf{w}\wedge\mathbf{u}}, \quad S_3 = \frac{i}{2} Q_{\mathbf{u}\wedge\mathbf{v}},$$

$$\widetilde{S}_1 = \frac{i}{2} \widetilde{Q}_{\mathbf{v}\wedge\mathbf{w}}, \quad \widetilde{S}_2 = \frac{i}{2} \widetilde{Q}_{\mathbf{w}\wedge\mathbf{u}}, \quad \widetilde{S}_3 = \frac{i}{2} \widetilde{Q}_{\mathbf{u}\wedge\mathbf{v}}.$$

One can verify that for these operators the above defined relations (5) and (6) are fulfilled and, hence, the internal symmetry in question provides realization of the idea of the Intrinsic Top and the concept of the Emergent Spin in the space of spin fields.

Equations of Spin Field

In what follows our main goal is to establish equations of the spin field and recognize their symmetry properties. To this end, let us consider natural general covariant differential operators of the first order for the spin field that are defined by the structure of the Intrinsic Top described above. These differential operators give a natural method to develop a theory of real spin fields outside the time (static theory). On this ground a dynamical theory of the real and complex spin fields is formulated as well.

Let ∇_i be the covariant derivative with respect to the connection belonging to the Riemann metric g_{ij}. The Christoffel symbols of this connection are

$$\Gamma^i_{jk} = \frac{1}{2} g^{il}(\partial_j g_{kl} + \partial_k g_{jl} - \partial_l g_{jk}).$$

The evident correspondence between v_i and ∇_i (v^i and $\nabla^i = g^{ij}\nabla_j$) make it possible to introduce two natural general covariant differential operators for the spin field

$$D = Q_\nabla, \quad \widetilde{D} = \widetilde{Q}_\nabla,$$

$$(D\mathbf{A})_{i_1 i_2 \cdots i_p} = p\nabla_{[i_1} a_{i_2 \cdots i_p]} - \nabla^l a_{l i_1 i_2 \cdots i_p},$$

$$(\widetilde{D}\mathbf{A})_{i_1 i_2 \cdots i_p} = (-1)^{p+1}(p\nabla_{[i_1} a_{i_2 \cdots i_p]} + \nabla^l a_{l i_1 i_2 \cdots i_p}).$$

These operators inherit the fundamental property of the operators $Q_\mathbf{v}$ and $\widetilde{Q}_\mathbf{v}$ since

$$D\widetilde{D} = \widetilde{D}D.$$

We conclude that the concept of Emergent Spin presupposes consideration of two dual general covariant equations

$$D\mathbf{A} = m\mathbf{A} \tag{10}$$

and

$$\widetilde{D}\mathbf{A} = -m\mathbf{A} \tag{11}$$

for the real spin field spatially extended, yet timeless.

Equations (10) and (11) are equivalent, and to show this, we introduce operators X and Y, which are analogous to the operator Z and are defined as follows:

$$X^{i_1 \cdots i_p}_{j_1 \cdots j_q} = 0, \quad \text{if} \quad p \neq q, \quad X^{i_1 \cdots i_p}_{j_1 \cdots j_p} = (-1)^{\frac{p(p-1)}{2}} \delta^{i_1 \cdots i_p}_{j_1 \cdots j_p},$$

$$Y^{i_1 \cdots i_p}_{j_1 \cdots j_q} = 0, \quad \text{if} \quad p \neq q, \quad Y^{i_1 \cdots i_p}_{j_1 \cdots j_p} = (-1)^{\frac{p(p+1)}{2}} \delta^{i_1 \cdots i_p}_{j_1 \cdots j_p}.$$

From the definition it follows that the operators X, Y, and Z commute with each other and the following relation are fulfilled:

$$XYZ = E, \quad X^2 = Y^2 = Z^2 = E, \quad \widetilde{Q}_\mathbf{v} = -YQ_\mathbf{v}Y, \quad \widetilde{Q}_\mathbf{v} = XQ_\mathbf{v}X,$$

$$\widetilde{D} = -YDY, \quad \widetilde{D} = XDX.$$

The last relations prove our statement and allow us to put forward the following conjecture. We take into account the operation of charge conjugation and define the operation of mass conjugation $m \to -m$ which means that if equation (10) describe a particle with mass m and a wave function $\mathbf{A}$, then equation (11) we put into correspondence to a particle with mass $-m$ and wave function $\widetilde{\mathbf{A}} = Y\mathbf{A}$.

To formulate the Lagrange formalism, let us define the positive definite fundamental bilinear quadratic form in the linear space in question as follows:

$$(\mathbf{A}|\mathbf{B}) = \sum_{p=0}^{n} \frac{1}{p!} a_{i_1\cdots i_p} b_{j_1\cdots j_p} g^{i_1 j_1} \cdots g^{i_p j_p}.$$

We have $(\mathbf{A}|\mathbf{B}) = (\mathbf{B}|\mathbf{A}), \quad (\mathbf{A}|\mathbf{A}) \geq 0$ and

$$(\mathbf{A}|Q_{\mathbf{v}}\mathbf{B}) = -(Q_{\mathbf{v}}\mathbf{A}|\mathbf{B}), \quad (\mathbf{A})|\widetilde{Q}_{\mathbf{v}}\mathbf{B}) = -(\widetilde{Q}_{\mathbf{v}}\mathbf{A}|\mathbf{B}),$$

$$(\mathbf{A}|Q_{\mathbf{v}\wedge\mathbf{w}}\mathbf{B}) = -(Q_{\mathbf{v}\wedge\mathbf{w}}\mathbf{A}|\mathbf{B}), \quad (\mathbf{A}|\widetilde{Q}_{\mathbf{v}\wedge\mathbf{w}}\mathbf{B})) = -(\widetilde{Q}_{\mathbf{v}\wedge\mathbf{w}}\mathbf{A}|\mathbf{B}),$$

We observe two identities

$$(\mathbf{A}|D\mathbf{B}) - (\mathbf{B}|D\mathbf{A}) = \nabla_i T^i, \quad (\mathbf{A}|\widetilde{D}\mathbf{B}) - (\mathbf{B}|\widetilde{D}\mathbf{A}) = \nabla_i \widetilde{T}^i. \tag{12}$$

The components of the vector fields T^i and $\widetilde{T}^i$ can be found from the relations

$$v_i T^i = (Q_{\mathbf{v}}\mathbf{A}|\mathbf{B}) = -(\mathbf{A}|Q_{\mathbf{v}}\mathbf{B}), \quad v_i \widetilde{T}^i = (\widetilde{Q}_{\mathbf{v}}\mathbf{A}|\mathbf{B}) = -(\mathbf{A}|\widetilde{Q}_{\mathbf{v}})\mathbf{B}).$$

Hence,

$$T^k = \sum_{p=0}^{n} \frac{1}{p!} (b_{i_1\cdots i_p} a^{ki_1\cdots i_p} - a_{i_1\cdots i_p} b^{ki_1\cdots i_p}),$$

$$\widetilde{T}^k = \sum_{p=0}^{n} \frac{(-1)^p}{p!} (b_{i_1\cdots i_p} a^{ki_1\cdots i_p} - a_{i_1\cdots i_p} b^{ki_1\cdots i_p}).$$

We see that the operators D and $\widetilde{D}$ are self-adjoint and $\nabla_i T^i = 0$ if $\mathbf{A}$ and $\mathbf{B}$ are solutions to the equation in question (and the same for $\nabla_i \widetilde{T}^i$). From (12) it follows that equations (10) and (11) can be derived from the dual Lagrangians

$$\pounds = \frac{1}{2}(\mathbf{A}|D\mathbf{A}) - \frac{m}{2}(\mathbf{A}|\mathbf{A}), \quad \widetilde{\pounds} = \frac{1}{2}(\mathbf{A}|\widetilde{D}\mathbf{A}) + \frac{m}{2}(\mathbf{A}|\mathbf{A}).$$

It is clear that $\pounds = 0$ if $\mathbf{A}$ is a solution of equation (10) and the same for $\widetilde{\pounds}$.

Now let us consider the symmetry properties of equations (10) and (11). Generally speaking, the operators $\widetilde{Q}_{\mathbf{v}\wedge\mathbf{w}\wedge\cdots\mathbf{z}}$ can commute with the operator D and the same holds for the pair $(\widetilde{D}, \quad Q_{\mathbf{v}\wedge\mathbf{w}\wedge\cdots\mathbf{z}})$. We keep in mind the characteristic relation $\widetilde{Q}_{\mathbf{v}}Q_{\mathbf{w}} = Q_{\mathbf{w}}\widetilde{Q}_{\mathbf{v}}$. It is evident that

$$D\widetilde{Q}_{\mathbf{v}} = \widetilde{Q}_{\mathbf{v}}D, \quad \widetilde{D}Q_{\mathbf{v}} = Q_{\mathbf{v}}\widetilde{D}$$

if

$$\nabla_i v_j = 0, \quad \Rightarrow R^m_{ijk}v_m = 0,$$

where R^m_{ijk} is the Riemann tensor of the connection Γ^i_{jk}, defined above. Now we consider the operators of the form $Q_{\mathbf{v}\wedge\mathbf{w}}$ and $\widetilde{Q}_{\mathbf{v}\wedge\mathbf{w}}$. The operators of this class are generated by the totally antisymmetric tensor field $v_iw_j - v_jw_i$ which is called simple. Hence, it is clear how to introduce the dual operators $Q_{\mathbf{S}}$ and $\widetilde{Q}_{\mathbf{S}}$ generated by an antisymmetric tensor field S_{ij} The component representation takes the form

$$(Q_{\mathbf{S}}\mathbf{A})_{i_1\cdots i_p} = \frac{p(p-1)}{2}S_{[i_1i_2}a_{i_3\cdots i_p]} - \frac{1}{2}S^{kl}a_{kli_1\cdots i_p} + pS_{k[i_1}a^k_{\cdot i_2\cdots i_p]},$$

$$(\widetilde{Q}_{\mathbf{S}}\mathbf{A})_{i_1\cdots i_p} = -\frac{p(p-1)}{2}S_{[i_1i_2}a_{i_3\cdots i_p]} + \frac{1}{2}S^{kl}a_{kli_1\cdots i_p} + pS_{k[i_1}a^k_{\cdot i_2\cdots i_p]}.$$

We have $D\widetilde{Q}_{\mathbf{S}} = \widetilde{Q}_{\mathbf{S}}D$, and $\widetilde{D}Q_{\mathbf{S}} = Q_{\mathbf{S}}\widetilde{D}$, if

$$\nabla_i S_{jk} = 0, \quad \Rightarrow R^m_{ijk}S_{ml} + R^m_{ijl}S_{km} = 0.$$

Thus, found is a deep connection between internal symmetry of the equations in question and the metric field $g_{ij}(x)$. Hence, this question requires a separate careful consideration and this will be done elsewhere.

We see that the rectilinear introduction of internal symmetry can not be realized. However, there is a very interesting exclusion here. When the number of the vector fields is equal to the dimension of the reference space we can consider instead of the $\mathbf{v}\wedge\mathbf{w}\cdots\wedge\mathbf{z}$ the antisymmetrical tensor $e_{i_1\cdots i_n}$. We set in this case

$$Q_{\mathbf{v}\wedge\mathbf{w}\cdots\wedge\mathbf{z}} = H, \quad \widetilde{Q}_{\mathbf{v}\wedge\mathbf{w}\cdots\wedge\mathbf{z}} = \widetilde{H}.$$

As above, for the operators H and $\widetilde{H}$ we have

$$D\widetilde{H} = \widetilde{H}D, \quad \widetilde{D}H = H\widetilde{D}$$

if $\nabla_k e_{i_1\cdots i_n} = 0$. This is the case if the essential component $e_{1\cdots n} = \sqrt{g}$. We write these dual operators in a component form for any dimension of the reference space

$$(H\mathbf{A})_{i_1\cdots i_p} = \frac{(-1)^n}{(n-p)!}(-1)^{\frac{(n-p)(n-p+1)}{2}} e_{i_1\cdots i_p j_1\cdots j_{n-p}} a^{j_1\cdots j_{n-p}},$$

$$(\widetilde{H}\mathbf{A})_{i_1\cdots i_p} = \frac{(-1)^n}{(n-p)!}(-1)^{\frac{p(p+1)}{2}} e_{i_1\cdots i_p j_1\cdots j_{n-p}} a^{j_1\cdots j_{n-p}}$$

and derive the following relations:

$$H^2 = \widetilde{H}^2 = (-1)^{\frac{n(n+1)}{2}} E, \quad HZ = (-1)^n ZH, \quad \widetilde{H}Z = (-1)^n Z\widetilde{H},$$

$$HQ_{\mathbf{v}} + (-1)^n Q_{\mathbf{v}} H = 0, \quad \widetilde{H}\widetilde{Q}_{\mathbf{v}} + (-1)^n \widetilde{Q}_{\mathbf{v}} \widetilde{H} = 0.$$

It is evident that

$$HD + (-1)^n DH = 0, \quad \widetilde{H}\widetilde{D} + (-1)^n \widetilde{D}\widetilde{H} = 0.$$

We also have $(\widetilde{H}H\mathbf{A})_{i_1\cdots i_p} = (-1)^{p(n+1)} a_{i_1\cdots i_p}$ and hence

$$\widetilde{H}H = H\widetilde{H} = Z$$

for the even dimension and

$$\widetilde{H}H = H\widetilde{H} = E$$

for the odd dimension. In the case $n = 3$ we have $H^2 = E, \quad \widetilde{H}^2 = E$.

It makes sense to write down equations (10) and (11) in the dimension $n = 3$. We set in this case $a = \alpha, \quad a_{ijk} = -e_{ijk}\beta, \quad a_{ij} = e_{ijk}b^k$, and using the formalism of the usual vector analysis in the general covariant form we get the following system of equations :

$$\begin{aligned} -\mathrm{div}\,\mathbf{a} &= m\,\alpha \\ -\mathrm{div}\,\mathbf{b} &= m\,\beta \\ \mathrm{rot}\,\mathbf{a} + \mathrm{grad}\,\beta &= m\,\mathbf{b} \\ \mathrm{rot}\,\mathbf{b} + \mathrm{grad}\,\alpha &= m\,\mathbf{a}. \end{aligned} \tag{13}$$

The dual equations takes the form

$$\begin{aligned} -\mathrm{div}\,\mathbf{a} &= m\,\alpha \\ -\mathrm{div}\,\mathbf{b} &= m\,\beta \\ -\mathrm{rot}\,\mathbf{a} + \mathrm{grad}\,\beta &= m\,\mathbf{b} \\ -\mathrm{rot}\,\mathbf{b} + \mathrm{grad}\,\alpha &= m\,\mathbf{a}. \end{aligned} \tag{14}$$

For $\widetilde{H}A = A$ we have $\alpha = \beta, \quad \mathbf{a} = \mathbf{b}$ and in the other case $\alpha = -\beta, \quad \mathbf{a} = -\mathbf{b}$. Eqs. (13) and (14) read:

$$\begin{aligned} -\mathrm{div}\,\mathbf{a} &= m\,\alpha \\ \mathrm{rot}\,\mathbf{a} + \mathrm{grad}\,\alpha &= m\,\mathbf{a}, \end{aligned}$$

$$\begin{aligned} -\mathrm{div}\,\mathbf{a} &= m\,\alpha \\ -\mathrm{rot}\,\mathbf{a} + \mathrm{grad}\,\alpha &= m\,\mathbf{a}. \end{aligned}$$

In what follows it will be useful to compare these equations with analogous equations in the dynamical theory of the spin field for better understanding of the dynamical theory itself and the role of the potential field $(f, \partial_i f)$ which represents the natural Time.

We supplement the equations (13) and (14) by the Einstein equations

$$G_{ij} = T_{ij}, \tag{15}$$

where T_{ij} can be found from the relation $\delta\mathcal{L} = \frac{1}{2}T_{ij}\delta g^{ij}$ or from the dual relation $\delta\widetilde{\mathcal{L}} = \frac{1}{2}T_{ij}\delta g^{ij}$. These equations have nontrivial solutions if $T_{ij} \neq 0$, since in the dimension $n = 3$ the curvature tensor $R^k_{ijl} = 0$, if $G_{ij} = 0$. From equations (10) and (11) it follows that $\mathbf{A}$ is an eigenvector of the self-adjoint operators with eigenvalue m. One can find eigenvectors and eigenvalues of the operators rot, D and $\widetilde{D}$ which belong to the metric field $g_{ij}(x)$. After Perelman's discovery it became possible to classify the set of 3-dimensional Riemann metric fields $g_{ij}(x)$ and to find solutions of the equations in question. This significant mathematical problem may have important physical applications since we need to know the laws of the initial states of the dynamical fields.

Bilateral Symmetry and Dynamical Equations

Bilateral symmetry is fundamental symmetry of nature defined by the gradient of Time. Hence, to formulate a dynamical theory of the spin field, we need first of all to recognize in what case a pair of spin fields $\mathbf{A}$ and $\overline{\mathbf{A}}$ has bilateral symmetry. Since the symmetry of right and left for the covector fields reads $\bar{v}_i = R_i^k v_k$, it is natural to start from the consideration of the operator

$$(R\mathbf{A})_{i_1 \cdots i_p} = R_{i_1}^{j_1} \cdots R_{i_p}^{j_p} a_{j_1 \cdots j_p} = (-1)^p (a_{i_1 \cdots i_p} - 2pt^k a_{k[i_2 \cdots i_p} t_{i_1]}),$$

$$(p = 0, 1, \cdots n)$$

that gives a straightforward representation of the bilateral symmetry. In this form the bilateral symmetry cannot be considered as an emergent property. But taking into account the definition of the operators E_{v}, I_{v} and Q_{v}, $\widetilde{Q}_{\mathrm{v}}$ we get sequentially

$$R = Z(E - 2E_{\mathrm{t}} I_{\mathrm{t}}) = -Q_{\mathrm{t}} \widetilde{Q}_{\mathrm{t}} = -\widetilde{Q}_{\mathrm{t}} Q_{\mathrm{t}}.$$

This operator representation shows that the bilateral symmetry is a property of the system in question (emergent property). We define that two spin fields $\mathbf{A}$ and $\overline{\mathbf{A}} = R\mathbf{A}$ possess the bilateral symmetry if

$$\overline{\mathbf{A}} = R\mathbf{A} = -Q_{\mathrm{t}} \widetilde{Q}_{\mathrm{t}} \mathbf{A}, \quad R^2 = E.$$

The operator R is invariant with respect to the transformation $\mathbf{t} \to -\mathbf{t}$.

However, the operator representation opens a new possibility to treatise the bilateral symmetry in the case of the spin field and introduce complex spin fields in accordance with the principle of sufficient cause: nothing happens without there being a reason why it should be thus rather than another outcome. Till now we have considered the real spin fields since there is no essential informal reason to introduce complex spin fields. Only the existence of these reasons has the foundational significance and represents real interest and motivation for the complexification. To introduce a complex spin field in accordance with the principle of sufficient cause and to conform that the bilateral symmetry is the fundamental symmetry of nature, we pay attention to the definition $i^2 = -1$. This gives

$$R == -Q_{\mathrm{t}} \widetilde{Q}_{\mathrm{t}} = i^2 Q_{\mathrm{t}} \widetilde{Q}_{\mathrm{t}} = (iQ_{\mathrm{t}})(i\widetilde{Q}_{\mathrm{t}}).$$

We conclude that for the complex spin fields

$$\mathbf{\Psi} = \mathbf{A} + i\mathbf{B}, \quad \overset{*}{\mathbf{\Psi}} = (\mathbf{A} - i\mathbf{B})$$

the bilateral symmetry can be defined as two different transformations of the form

$$\overline{\mathbf{\Psi}} = R\mathbf{\Psi} = iQ_{\mathbf{t}}\mathbf{\Psi}, \quad R^2 = E$$

and

$$\overline{\mathbf{\Psi}} = R\mathbf{\Psi} = i\widetilde{Q}_{\mathbf{t}}\mathbf{\Psi}, \quad R^2 = E.$$

In view of this we at first formulate a dynamical theory of real spin field and after that consider a dynamical theory of a complex spin field which is introduced here in accordance with the principle of sufficient cause. This presupposes as well that at the first stage we derive dynamical equations of the spin field in the operator form (using the formalism of the auxiliary metric). After that we write down these equations in a new and important (from a definite point of view) representation which is defined by the main differential operators of the Maxwell equations written in the general covariant and four-dimensional form [2]. To this end, let us consider the bilinear quadratic form of the spin fields that is defined by the auxiliary metric $\overline{g}_{ij} = 2t_i t_j - g_{ij}, \quad \overline{g}^{ij} = 2t^i t^j - g^{ij}$,

$$\langle \mathbf{A}|\mathbf{B}\rangle = \sum\nolimits_{p=0}^{n} \frac{1}{p!} a_{i_1\cdots i_p} b_{j_1\cdots j_p} \overline{g}^{i_1 j_1} \cdots \overline{g}^{i_p j_p}. \tag{16}$$

Since $\overline{g}^{ij} = R^j_k g^{ik}$, we have the following relation between two quadratic forms:

$$\langle \mathbf{A}|\mathbf{B}\rangle = (\mathbf{A}|R\mathbf{B}) = -(\mathbf{A}|Q_{\mathbf{t}}\widetilde{Q}_{\mathbf{t}}\mathbf{B}) = (Q_{\mathbf{t}}\mathbf{A}|\widetilde{Q}_{\mathbf{t}}\mathbf{B}) = (\widetilde{Q}_{\mathbf{t}}\mathbf{A}|Q_{\mathbf{t}}\mathbf{B}). \tag{17}$$

We introduce also the operators $L_{\mathbf{v}}$ and $\widetilde{L}_{\mathbf{v}}$ that correspond to $Q_{\mathbf{v}}$ and $\widetilde{Q}_{\mathbf{v}}$ but are defined by the auxiliary metric. We have the following relations between these operators:

$$L_{\mathbf{v}} = \widetilde{Q}_{\mathbf{v}}Z + (\mathbf{t},\mathbf{v})(Q_{\mathbf{t}} - \widetilde{Q}_{\mathbf{t}}Z) \quad \widetilde{L}_{\mathbf{v}} = Q_{\mathbf{v}}Z + (\mathbf{t},\mathbf{v})(\widetilde{Q}_{\mathbf{t}} - Q_{\mathbf{t}}Z).$$

It is easy to see that

$$L_{\mathbf{t}} = Q_{\mathbf{t}}, \quad \widetilde{L}_{\mathbf{t}} = \widetilde{Q}_{\mathbf{t}} \quad \text{and} \quad L_{\mathbf{v}} = \widetilde{Q}_{\mathbf{v}}Z, \quad \widetilde{L}_{\mathbf{v}} = Q_{\mathbf{v}}Z,$$

when a vector $\mathbf{v}$ is orthogonal to the stream of time, $(\mathbf{t}, \mathbf{v}) = 0$.

Now we define the main differential operators of the dynamical theory of the spin field. Let $\overline{\Gamma}^i_{jk}$ be the Christoffel symbols belonging to the auxiliary metric $\overline{g}_{ij} = 2t_i t_j - g_{ij}$ and $\overline{\nabla}_i$ be the covariant derivative with respect to this auxiliary connection. We can consider the dual linear differential operators of the first order $L_{\overline{\nabla}}$ and $\widetilde{L}_{\overline{\nabla}}$ but there is a simpler and more elegant approach to derive equations of the real and complex spin fields. We introduce the connection

$$\breve{\Gamma}^i_{jk} = \Gamma^i_{jk} + t^i \nabla_j t_k - t_k \nabla_j t^i$$

with the following remarkable properties:

$$\breve{\nabla}_i t_j = 0, \quad \breve{\nabla}_i g_{jk} = 0,$$

where $\breve{\nabla}_i$ is the covariant derivative with respect to the connection $\breve{\Gamma}^i_{jk}$. We can say that the gradient of Time and the metric field are covariantly constant with respect to this connection. We define the main differential operators Π and $\widetilde{\Pi}$ of the dynamical theory of the spin field as follows:

$$\Pi = L_{\breve{\nabla}}, \quad \widetilde{\Pi} = \widetilde{L}_{\breve{\nabla}},$$

where (in accordance with the definition)

$$(\Pi\mathbf{A})_{i_1\cdots i_p} = p\breve{\nabla}_{[i_1} a_{i_2\cdots i_p]} + \breve{\nabla}^k a_{k i_1\cdots i_p} - 2t^k \breve{\nabla}_{\mathbf{t}} a_{k i_1\cdots i_p}, \quad p = 0, 1, 2, 3, 4,$$

$$(\widetilde{\Pi}\mathbf{A})_{i_1\cdots i_p} = (-1)^{p+1}(p\breve{\nabla}_{[i_1} a_{i_2\cdots i_p]} - \breve{\nabla}^k a_{k i_1\cdots i_p} + 2t^k \breve{\nabla}_{\mathbf{t}} a_{k i_1\cdots i_p}),$$

since $\overline{g}^{kl}\breve{\nabla}_l = 2t^k \breve{\nabla}_{\mathbf{t}} - \breve{\nabla}^k$.

Following the main idea of the concept of the Emergent Spin, we consider the dual Lagrangians

$$\mathcal{L}_{\mathbf{t}} = \frac{1}{2}\langle \mathbf{A}|\Pi\mathbf{A}\rangle - \frac{m}{2}\langle \mathbf{A}|\mathbf{A}\rangle.$$

$$\widetilde{\mathcal{L}}_{\mathbf{t}} = \frac{1}{2}\langle \mathbf{A}|\widetilde{\Pi}\mathbf{A}\rangle + \frac{m}{2}\langle \mathbf{A}|\mathbf{A}\rangle.$$

Since for any vector field

$$\breve{\nabla}_i T^i = \nabla_i T^i - \varphi\, t_i T^i, \quad \varphi = \nabla_i t^i,$$

the basic identities (12) takes the form

$$\langle \mathbf{A}|\Pi\mathbf{B} + \frac{1}{2}Q_{\mathbf{t}}\varphi\mathbf{B}\rangle - \langle \mathbf{B}|\Pi\mathbf{A} + \frac{1}{2}Q_{\mathbf{t}}\varphi\mathbf{A}\rangle = \nabla_k T^k.$$

$$\langle \mathbf{A}|\widetilde{\Pi}\mathbf{B} + \frac{1}{2}\widetilde{Q}_{\mathbf{t}}\varphi\mathbf{B}\rangle - \langle \mathbf{B}|\widetilde{\Pi}\mathbf{A} + \frac{1}{2}\widetilde{Q}_{\mathbf{t}}\varphi\mathbf{A}\rangle = \nabla_k \widetilde{T}^k,$$

where T^k and $\widetilde{T}^k$ are defined from the relations:

$$v_k T^k = \langle L_{\mathbf{v}}\mathbf{A}|\mathbf{B}\rangle, \quad v_k \widetilde{T}^k = \langle \widetilde{L}_{\mathbf{v}}\mathbf{A}|\mathbf{B}\rangle.$$

The dual dynamical equations for the real spin field read:

$$\Pi\mathbf{A} + \frac{1}{2}\varphi Q_{\mathbf{t}}\mathbf{A} = m\mathbf{A}, \tag{18}$$

$$\widetilde{\Pi}\mathbf{A} + \frac{1}{2}\varphi\widetilde{Q}_{\mathbf{t}}\mathbf{A} = -m\mathbf{A}. \tag{19}$$

The Lagrangians in question are invariant with respect to time reversal T : $\mathbf{t} \to -\mathbf{t}$ and since

$$\Pi\widetilde{Q}_{\mathbf{t}} - \widetilde{Q}_{\mathbf{t}}\Pi = 0, \quad \langle \widetilde{Q}_{\mathbf{t}}\mathbf{A}|\mathbf{B}\rangle = -\langle \mathbf{A}|\widetilde{Q}_{\mathbf{t}}\mathbf{B}\rangle,$$

$$\widetilde{\Pi}Q_{\mathbf{t}} - Q_{\mathbf{t}}\widetilde{\Pi} = 0, \quad \langle Q_{\mathbf{t}}\mathbf{A}|\mathbf{B}\rangle = -\langle \mathbf{A}|Q_{\mathbf{t}}\mathbf{B}\rangle,$$

they are invariant with respect to the transformations

$$\overline{\mathbf{A}} = \frac{\Lambda}{|\Lambda|}\mathbf{A}, \quad \Lambda = \Lambda_1 E + \Lambda_2\widetilde{Q}_{\mathbf{t}}, \quad |\Lambda|^2 = {\Lambda_1}^2 + {\Lambda_2}^2,$$

$$\overline{\mathbf{A}} = \frac{\Lambda}{|\Lambda|}\mathbf{A}, \quad \Lambda = \Lambda_1 E + \Lambda_2 Q_{\mathbf{t}}, \quad |\Lambda|^2 = {\Lambda_1}^2 + {\Lambda_2}^2.$$

Setting in our modified dual identities $\mathbf{B} = \widetilde{Q}_{\mathbf{t}}\mathbf{A}$ and $\mathbf{B} = Q_{\mathbf{t}}\mathbf{A}$, we conclude that

$$\nabla_k C^k = 0, \quad \nabla_k \widetilde{C}^k = 0,$$

where C^k and $\widetilde{C}^k$ are the components of the vectors which can be found from the relations

$$v_k C^k = \langle L_{\mathbf{v}}\mathbf{A}|\widetilde{Q}_{\mathbf{t}}\mathbf{A}\rangle, \quad v_k \widetilde{C}^k = \langle \widetilde{L}_{\mathbf{v}}\mathbf{A}|Q_{\mathbf{t}}\mathbf{A}\rangle. \tag{20}$$

The dual laws of conservations in question are tightly connected with the gradient of Time and on this ground we consider the operators $Q_{\mathbf{t}}$ and $\widetilde{Q}_{\mathbf{t}}$ as dual operators of the electric charge. These operators ensure the existence of the probability measure. Let us prove this statement. To find the dual charge densities and the dual physical currents, we set

$$C^k = t^k(\mathbf{t}, \mathbf{C}) + C^k - t^k(\mathbf{t}, \mathbf{C}) = \rho t^k + J^k,$$

$$\widetilde{C}^k = t^k(\mathbf{t}, \widetilde{\mathbf{C}}) + \widetilde{C}^k - t^k(\mathbf{t}, \widetilde{\mathbf{C}}) = \widetilde{\rho} t^k + \widetilde{J}^k.$$

Thus, the dual laws of charge conservation read:

$$D_{\mathbf{t}}(\sqrt{g}\rho) + \partial_k(\sqrt{g}J^k) = 0, \quad t_k J^k = 0.$$

$$D_{\mathbf{t}}(\sqrt{g}\widetilde{\rho} + \partial_k(\sqrt{g}\widetilde{J}^k) = 0, \quad t_k \widetilde{J}^k = 0.$$

Since $\rho = t_k C^k, \quad \widetilde{\rho} = t_k \widetilde{C}^k$, we have from (20)

$$\rho = t_k C^k = \langle Q_{\mathbf{t}}\mathbf{A} | \widetilde{Q}_{\mathbf{t}}\mathbf{A}\rangle = (Q_{\mathbf{t}}Q_{\mathbf{t}}\mathbf{A} | \widetilde{Q}_{\mathbf{t}}\widetilde{Q}_{\mathbf{t}}\mathbf{A}) = (\mathbf{A}|\mathbf{A}) = \widetilde{\rho}.$$

We have arrived at the needed fundamental result because $\rho = (\mathbf{A}|\mathbf{A}) > 0$ and $(\mathbf{A}|\mathbf{A}) = 0$ provides that $\mathbf{A} = 0$.

The importance of this result can be understood as follows. An experimentalist cannot observe a quantum system without producing a serious disturbance and hence he cannot expect to find any causal relation between the results of his observations. However, causality will still be assumed to apply to undisturbed systems and the equations which will be set up to describe an undisturbed system will be differential equations expressing a causal connection between the conditions at one time and the conditions at later time. The deterministic equations of this kind should guarantee the existence of the probabilistic measure in the space of solutions, which is invariant with respect to time translation. This measure has the foundational meaning since it opens up the possibilities of probabilistic description when signals from the disturbed system may be non-predictable; however, their statistical properties are reproduced with time. For example, a similar situation occurs in the theory of turbulence where the methods of the probability theory have the strict and quite practical meaning. It is evident that the same situation holds for the so-called quantum systems.

Further, in the definition of the dual operators H and $\widetilde{H}$ we substitute instead of the Riemann metric field $g_{ij}(x)$ the auxiliary metric field $\overline{g}_{ij}(x)$ and hence we get the dual operators

$$J = -HQ_{\mathrm{t}}\widetilde{Q}_{\mathrm{t}} = Q_{\mathrm{t}}H\widetilde{Q}_{\mathrm{t}}, \quad \widetilde{J} = -\widetilde{H}Q_{\mathrm{t}}\widetilde{Q}_{\mathrm{t}} = \widetilde{Q}_{\mathrm{t}}\widetilde{H}Q_{\mathrm{t}},$$

which together with the operators Q_{t} and $\widetilde{Q}_{\mathrm{t}}$ act in the space of solutions of the corresponding dual equations. The following relations

$$J^2 = \widetilde{J}^2 = (-1)^{\frac{(n+1)(n+2)}{2}} E$$

give for the case $n = 4$

$$J^2 = -E, \quad \widetilde{J}^2 = -E, \quad JQ_{\mathrm{t}} + Q_{\mathrm{t}}J = 0, \quad \widetilde{J}\widetilde{Q}_{\mathrm{t}} + \widetilde{Q}_{\mathrm{t}}\widetilde{J} = 0.$$

To complete our consideration, we need to prove that $\breve{\nabla}_i e_{jklm} = 0$. We have in any dimension

$$\breve{\nabla}_k e_{i_1\cdots i_n} = \nabla_k e_{i_1\cdots i_n} - T^l_{ki_1} e_{li_2\cdots i_n} - \cdots - T^l_{ki_n} e_{i_1\cdots i_{n-1}l},$$

where $T^i_{jk} = t^i\nabla_j t_k - t_k\nabla_j t^i$. Since $\nabla_k e_{i_1\cdots i_n} = 0$, then our result follows from the identity

$$(n+1)T^m_{k[l}e_{i_1\cdots i_n]} = T^m_{kl}e_{i_1\cdots i_n} - nT^m_{k[i_1}e_{|l|i_2\cdots i_n]}, \quad T^l_{kl} = 0.$$

Hence, the commutation relations hold valid

$$\Pi\widetilde{J} - \widetilde{J}\Pi = 0, \quad \widetilde{\Pi}J - J\widetilde{\Pi} = 0,$$

however, the Lagrangians in question are not invariant with respect to the transformations

$$\overline{\mathbf{A}} = \frac{\Lambda}{|\Lambda|}\mathbf{A}, \quad \Lambda = \Lambda_1 E + \Lambda_2\widetilde{J}, \quad |\Lambda|^2 = {\Lambda_1}^2 + {\Lambda_2}^2,$$

$$\overline{\mathbf{A}} = \frac{\Lambda}{|\Lambda|}\mathbf{A}, \quad \Lambda = \Lambda_1 E + \Lambda_2 J, \quad |\Lambda|^2 = {\Lambda_1}^2 + {\Lambda_2}^2.$$

On this reason, the vector field T^k at $\mathbf{B} = \widetilde{J}\mathbf{A}$ should be trivial. Let us prove this statement. We have

$$v_i T^i = \langle L_{\mathbf{v}}\mathbf{A}|\widetilde{J}\mathbf{A}\rangle = -\langle \mathbf{A}|\widetilde{J}L_{\mathbf{v}}\mathbf{A}\rangle = -\langle L_{\mathbf{v}}\mathbf{A}|\widetilde{J}\mathbf{A}\rangle = -v_i T^i.$$

Since $v_i T^i = 0$ for any vector field v_i, then $T^i = 0$.

Let A_i be components of the potential $\mathbf{a}$ of the electromagnetic field. In accordance with (20), we write the dual Lagrangians of interactions of the electromagnetic field and the real spin field:

$$\mathcal{L}_{int} = -\frac{1}{2} q\, A_i C^i = -\frac{1}{2} q \langle L_{\mathbf{a}} \mathbf{A} | \widetilde{Q}_{\mathbf{t}} \mathbf{A} \rangle,$$

$$\widetilde{\mathcal{L}}_{int} = -\frac{1}{2} q\, A_i \widetilde{C}^i = -\frac{1}{2} q \langle \widetilde{L}_{\mathbf{a}} \mathbf{A} | Q_{\mathbf{t}} \mathbf{A} \rangle,$$

where q is a constant of interaction. Taking into account the electromagnetic interactions, we write the dual equations of the real spin field

$$(\Pi + \frac{1}{2} \varphi Q_{\mathbf{t}}) \mathbf{A} + q L_{\mathbf{a}} \widetilde{Q}_{\mathbf{t}} \mathbf{A} = m \mathbf{A}, \tag{21}$$

$$(\widetilde{\Pi} + \frac{1}{2} \varphi \widetilde{Q}_{\mathbf{t}}) \mathbf{A} + q \widetilde{L}_{\mathbf{a}} Q_{\mathbf{t}} \mathbf{A} = -m \mathbf{A}. \tag{22}$$

The real spin field can not be an eigenvector of the dual operators of the electric charge since

$$Q_{\mathbf{t}}^2 = -E, \quad \widetilde{Q}_{\mathbf{t}}^2 = -E$$

and hence, its eigenvalues equal $\pm i$. For a complex spin field

$$\mathbf{\Psi} = \mathbf{A} + i \mathbf{B},$$

we can consider the dual equations

$$(\Pi + \frac{1}{2} \varphi Q_{\mathbf{t}}) \mathbf{\Psi} + q L_{\mathbf{a}} \widetilde{Q}_{\mathbf{t}} \mathbf{\Psi} = m \mathbf{\Psi}, \tag{23}$$

$$(\widetilde{\Pi} + \frac{1}{2} \varphi \widetilde{Q}_{\mathbf{t}}) \mathbf{\Psi} + q \widetilde{L}_{\mathbf{a}} Q_{\mathbf{t}} \mathbf{\Psi} = -m \mathbf{\Psi} \tag{24}$$

under the conditions

$$\widetilde{Q}_{\mathbf{t}} \mathbf{\Psi} = -i \mathbf{\Psi}, \quad Q_{\mathbf{t}} \mathbf{\Psi} = -i \mathbf{\Psi},$$

which provide conservation of the number of degrees of freedom of the real spin field.

Now we represent eqs. (21) and (22) in a promised impressible form using the results obtained under the derivation of the really Maxwell equations in the evidently dynamical and general covariant form [2]. First of all, remind some definitions of the vector algebra and vector analysis in the four-dimensional and general covariant form [2]. The operator rot is defined for the vector fields as follows:

$$(\mathrm{rot}\mathbf{M})^i = e^{ijkl} t_j \partial_k \mathrm{M}_l = \frac{1}{2} e^{ijkl} t_j (\partial_k \mathrm{M}_l - \partial_l \mathrm{M}_k).$$

It is easy to show that

$$(\mathbf{M}|\,\mathrm{rot}\mathbf{N}) + \mathrm{div}[\mathbf{M} \times \mathbf{N}] = (\mathrm{rot}\mathbf{M}|\,\mathbf{N}),$$

where

$$[\mathbf{M} \times \mathbf{N}]^i = e^{ijkl} t_j \mathrm{M}_k \mathrm{N}_l$$

is the vector product of two vector fields $\mathbf{M}$ and $\mathbf{N}$, $\mathrm{div}\,\mathbf{M} = \nabla_i \mathrm{M}^i$. Thus, the operator rot is self-adjoint. To find eigenvectors and eigenvalues of this operator, one should consider the equation $\mathrm{rot}\,\mathbf{M} = \sigma \mathbf{M}$. We also mention that

$$(\mathrm{grad}\,\varphi)_i = \triangle_i \varphi, \quad \triangle_i = \nabla_i - t_i \nabla_{\mathbf{t}}$$

and

$$\mathrm{rot}\,\mathrm{grad} = 0, \quad \mathrm{div}\,\mathrm{rot} = 0.$$

To apply formalisms of the vector analysis to equation (21), we consider the mapping of the real spin field

$$\mathbf{A} = (a, a_i, a_{ij}, a_{ijk}, a_{ijkl})$$

onto two scalars κ and μ, two pseudo-scalars λ and ν, two vectors $\mathbf{K}$ and $\mathbf{M}$, two pseudo-vectors $\mathbf{L}$ and $\mathbf{N}$ all orthogonal to the stream of time,

$$(\mathbf{t}, \mathbf{K}) = (\mathbf{t}, \mathbf{L}) = (\mathbf{t}, \mathbf{M}) = (\mathbf{t}, \mathbf{N}) = 0.$$

This mapping is defined as follows:

$$\kappa = t^i a_i, \quad \mu = a, \quad \lambda = \frac{1}{3!} t_l e^{ijkl} a_{ijk}, \quad \nu = \frac{1}{4!} e^{ijkl} a_{ijkl},$$

$$\mathrm{K}^i = h^i_m a^m, \quad \mathrm{M}^i = t_k a^{ki}, \quad \mathrm{L}^i = \frac{1}{3!} h^i_m e^{mjkl} a_{jkl}, \quad \mathrm{N}^i = t_k \tilde{a}^{ki},$$

where $h^i_j = \delta^i_j - t^i t_j, \quad \widetilde{a}^{ki} = \frac{1}{2} e^{kijl} a_{jl}.$

The inverse mapping has the form

$$a = \mu, \quad a_i = \mathrm{K}_i + \kappa t_i, \quad a_{ij} = t_i \mathrm{M}_j - t_j \mathrm{M}_i + e_{ijkl} t^k \mathrm{N}^l,$$

$$a_{ijk} = e_{mijk} \mathrm{L}^m + t^m e_{ijkm} \lambda, \quad a_{ijkl} = e_{ijkl}\, \nu.$$

Further, we write down equation (21) in a component form setting $p = 0, 1, 2, 3, 4$ and after some transformations arrive at the following symmetric system of equations which involves four scalar and four vector equations:

$$\begin{array}{l} (\nabla_{\mathbf{t}} + \frac{1}{2}\varphi)\mu - q\phi\,\kappa = \nabla_i \mathrm{M}^i - q\Phi_i \mathrm{K}^i + m\,\kappa \\ (\nabla_{\mathbf{t}} + \frac{1}{2}\varphi)\kappa + q\phi\,\mu = \nabla_i \mathrm{K}^i + q\Phi_i \mathrm{M}^i - m\,\mu \\ (\nabla_{\mathbf{t}} + \frac{1}{2}\varphi)\lambda - q\phi\,\nu = \nabla_i \mathrm{L}^i - q\Phi_i \mathrm{N}^i - m\,\nu \\ (\nabla_{\mathbf{t}} + \frac{1}{2}\varphi)\nu + q\phi\,\lambda = \nabla_i \mathrm{N}^i + q\Phi_i \mathrm{L}^i + m\,\lambda \end{array} \tag{25}$$

$$\begin{array}{l} (\nabla_{\mathbf{t}} + \frac{1}{2}\varphi)\mathrm{M}_i - q\phi \mathrm{K}_i = (\mathrm{rot}\,\mathbf{N})_i + q[\mathbf{\Phi} \times \mathbf{L}]_i + \triangle_i\,\mu - q\Phi_i\,\kappa - m\,\mathrm{K}_i \\ (\nabla_{\mathbf{t}} + \frac{1}{2}\varphi)\mathrm{K}_i + q\phi \mathrm{M}_i = -(\mathrm{rot}\,\mathbf{L})_i + q[\mathbf{\Phi} \times \mathbf{N}]_i + \triangle_i\,\kappa + q\Phi_i\,\mu + m\,\mathrm{M}_i \\ (\nabla_{\mathbf{t}} + \frac{1}{2}\varphi)\mathrm{L}_i - q\phi \mathrm{N}_i = (\mathrm{rot}\,\mathbf{K})_i + q[\mathbf{\Phi} \times \mathbf{M}]_i + \triangle_i\,\lambda - q\Phi_i\,\nu + m\,\mathrm{N}_i \\ (\nabla_{\mathbf{t}} + \frac{1}{2}\varphi)\mathrm{N}_i + q\phi \mathrm{L}_i = -(\mathrm{rot}\,\mathbf{M})_i + q[\mathbf{\Phi} \times \mathbf{K}]_i + \triangle_i\,\nu + q\Phi_i\,\lambda - m\,\mathrm{L}_i \end{array} \tag{26}$$

where

$$\nabla_{\mathbf{t}} = t^i \nabla_i, \quad \varphi = \nabla_i t^i \quad \text{and} \quad \phi = t^i A_i$$

is a scalar potential of the electromagnetic field and

$$\Phi_i = A_i - \phi t_i, \quad t^i \Phi_i = 0$$

is its vector potential. Under the gauge transformation $A_i \Rightarrow A_i + \partial_i \Lambda$ the scalar and vector potentials transform as follows:

$$\phi' = \phi + D_{\mathbf{t}}\Lambda, \quad \Phi_i{}' = \Phi_i + \triangle_i \Lambda, \quad \triangle_i = \nabla_i - t_i \nabla_{\mathbf{t}}.$$

We also write our fundamental equations in the invariant form:

$$\begin{array}{l} (\nabla_{\mathbf{t}} + \frac{1}{2}\varphi)\mu - q\phi\,\kappa = \mathrm{div}\,\mathbf{M} - q(\mathbf{\Phi}|\mathbf{K}) + m\,\kappa \\ (\nabla_{\mathbf{t}} + \frac{1}{2}\varphi)\kappa + q\phi\,\mu = \mathrm{div}\,\mathbf{K} + q(\mathbf{\Phi}|\mathbf{M}) - m\,\mu \\ (\nabla_{\mathbf{t}} + \frac{1}{2}\varphi)\lambda - q\phi\,\nu = \mathrm{div}\,\mathbf{L} - q(\mathbf{\Phi}|\mathbf{N}) - m\,\nu \\ (\nabla_{\mathbf{t}} + \frac{1}{2}\varphi)\nu + q\phi\,\lambda = \mathrm{div}\,\mathbf{N} + q(\mathbf{\Phi}|\mathbf{L}) + m\,\lambda \end{array} \tag{27}$$

$$\begin{aligned}(\nabla_{\mathbf{t}} + \tfrac{1}{2}\varphi)\mathbf{M} - q\phi\mathbf{K} &= (\mathrm{rot}\,\mathbf{N}) + q[\mathbf{\Phi}\times\mathbf{L}] + \mathrm{grad}\,\mu - q\mathbf{\Phi}\,\kappa - m\,\mathbf{K}\\ (\nabla_{\mathbf{t}} + \tfrac{1}{2}\varphi)\mathbf{K} + q\phi\mathbf{M} &= -(\mathrm{rot}\,\mathbf{L}) + q[\mathbf{\Phi}\times\mathbf{N}] + \mathrm{grad}\,\kappa + q\mathbf{\Phi}\,\mu + m\,\mathbf{M}\\ (\nabla_{\mathbf{t}} + \tfrac{1}{2}\varphi)\mathbf{L} - q\phi\mathbf{N} &= \mathrm{rot}\,\mathbf{K} + q[\mathbf{\Phi}\times\mathbf{M}] + \mathrm{grad}\,\lambda - q\mathbf{\Phi}\,\nu + m\,\mathbf{N}\\ (\nabla_{\mathbf{t}} + \tfrac{1}{2}\varphi)\mathbf{N} + q\phi\mathbf{L} &= -\mathrm{rot}\,\mathbf{M} + q[\mathbf{\Phi}\times\mathbf{K}] + \mathrm{grad}\,\nu + q\mathbf{\Phi}\,\lambda - m\,\mathbf{L}\end{aligned} \tag{28}$$

For the complex spin field the equation $\widetilde{Q}_{\mathbf{t}}\Psi = -i\Psi$ has the solution

$$\kappa = i\mu, \quad \nu = i\lambda, \quad \mathbf{K} = i\mathbf{M}, \quad \mathbf{N} = i\mathbf{L},$$

where μ, ν, κ, λ and $\mathbf{K}$, $\mathbf{L}$, $\mathbf{M}$, $\mathbf{N}$ are complex.

In this case, our equations read:

$$\begin{aligned}(\nabla_{\mathbf{t}} + \tfrac{1}{2}\varphi - iq\phi)\mu &= (\nabla - iq\mathbf{\Phi}|\mathbf{M}) + im\,\mu\\ (\nabla_{\mathbf{t}} + \tfrac{1}{2}\varphi - iq\phi)\lambda &= (\nabla - iq\mathbf{\Phi}|\mathbf{L}) - im\,\lambda\end{aligned} \tag{29}$$

$$\begin{aligned}(\nabla_{\mathbf{t}} + \tfrac{1}{2}\varphi - iq\phi)\mathbf{L} &= i[(\nabla - iq\mathbf{\Phi})\times\mathbf{M}] + (\triangle - iq\mathbf{\Phi})\,\lambda + im\,\mathbf{L}\\ (\nabla_{\mathbf{t}} + \tfrac{1}{2}\varphi - iq\phi)\mathbf{M} &= i[(\nabla - iq\mathbf{\Phi})\times\mathbf{L}] + (\triangle - iq\mathbf{\Phi})\,\mu - im\,\mathbf{M}\end{aligned} \tag{30}$$

and we can quite succesfully apply these eqs. in the case $g_{ij} = \delta_{ij}, \quad t = a_i x^i$ to an exact solution of the Coulomb problem in the hydrogen atom. To this end we need to use the vector spherical harmonics $\mathbf{Y}^L_{JM}(\theta,\varphi)$ [5] and discover for their contravariant cyclic components the following relations:

$$\sqrt{\frac{(J-M+1)(2J+3)}{J+1}}\mathbf{Y}^{J+1}_{JM}(\theta,\varphi) - \sqrt{\frac{(J+M+1)(J+2)}{J+1}}\mathbf{Y}^{J+1}_{J+1M}(\theta,\varphi) =$$

$$[\sqrt{2(J-M+2)}Y_{J+1M-1}(\theta,\varphi)]\mathbf{e}_{+1} - [\sqrt{J+M+1}Y_{J+1M}(\theta,\varphi)]\mathbf{e}_0,$$

$$\sqrt{\frac{(J+M)(2J-1)}{J}}\mathbf{Y}^{J-1}_{JM}(\theta,\varphi) - \sqrt{\frac{(J-M)(J-1)}{J}}\mathbf{Y}^{J-1}_{J-1M}(\theta,\varphi) =$$

$$[\sqrt{2(J+M-1)}Y_{J-1M-1}(\theta,\varphi)]\mathbf{e}_{+1} + [\sqrt{J-M}Y_{J-1M}(\theta,\varphi)]\mathbf{e}_0.$$

BILATERAL SYMMETRY AND NEUTRINO CHARGE

The fundamental dual Lagrangians of the spin field which are associated with the new representation of bilateral symmetry take the form

$$\mathcal{L}_{\mathrm{t}} = -\frac{i}{2}\langle \widetilde{Q}_{\mathrm{t}} \overset{*}{\mathbf{\Psi}} | \Pi \mathbf{\Psi} \rangle + \frac{i}{2}\langle \widetilde{Q}_{\mathrm{t}} \mathbf{\Psi} | \Pi \overset{*}{\mathbf{\Psi}} \rangle + im\langle \widetilde{Q}_{\mathrm{t}} \overset{*}{\mathbf{\Psi}} | \mathbf{\Psi} \rangle,$$

$$\widetilde{\mathcal{L}}_{\mathrm{t}} = -\frac{i}{2}\langle Q_{\mathrm{t}} \overset{*}{\mathbf{\Psi}} | \widetilde{\Pi} \mathbf{\Psi} \rangle + \frac{i}{2}\langle Q_{\mathrm{t}} \mathbf{\Psi} | \widetilde{\Pi} \overset{*}{\mathbf{\Psi}} \rangle + im\langle Q_{\mathrm{t}} \overset{*}{\mathbf{\Psi}} | \mathbf{\Psi} \rangle.$$

The relations

$$\langle \widetilde{Q}_{\mathrm{t}} \overset{*}{\mathbf{\Psi}} | \mathbf{\Psi} \rangle = (Q_{\mathrm{t}} \overset{*}{\mathbf{\Psi}} | \mathbf{\Psi}), \quad \langle Q_{\mathrm{t}} \overset{*}{\mathbf{\Psi}} | \mathbf{\Psi} \rangle = (\widetilde{Q}_{\mathrm{t}} \overset{*}{\mathbf{\Psi}} | \mathbf{\Psi})$$

clearly demonstrate the role of the new representation of the bilateral symmetry since

$$(Q_{\mathrm{t}} \mathbf{A} | \mathbf{A}) = (\widetilde{Q}_{\mathrm{t}} \mathbf{A} | \mathbf{A}) = 0$$

identically. We see that the Lagrangians in question are not invariant with respect to the Time reversal $T: \quad t_i \to -t_i$, (they change the sign, $\mathcal{L}_{-\mathrm{t}} = -\mathcal{L}_{\mathrm{t}}, \quad \widetilde{\mathcal{L}}_{-\mathrm{t}} = -\widetilde{\mathcal{L}}_{\mathrm{t}}$). We conclude that the energy-momentum tensor (that can be derived from the equations $\delta \mathcal{L}_{\mathrm{t}} = \frac{1}{2} T_{ij} \delta \overline{g}^{ij}, \quad \delta \widetilde{\mathcal{L}}_{\mathrm{t}} = \frac{1}{2} T_{ij} \delta \widetilde{\overline{g}}^{ij}$) changes sign under the Time reversal. Thus, the equations of gravidynamics [2]

$$G_{ij} + T_{ij} = \varepsilon t_i t_j$$

in the case of the complex spin field are not invariant under the Time reversal. This means that the gravitational interactions of the complex spin fields break invariance with respect to the Time reversal and with this we at first observe some rational visible evidence of the asymmetry between matter and antimatter.

Now we prove the existence of the probability measure in this new situation. Our dual Lagrangians are invariant with respect to the transformations

$$\mathbf{\Psi} \Rightarrow \frac{\Lambda}{|\Lambda|} \mathbf{\Psi}, \quad \Lambda = \Lambda_1 + \Lambda_2 \, i, \quad |\Lambda|^2 = {\Lambda_1}^2 + {\Lambda_2}^2$$

and as a consequence we arrive at the equations

$$\nabla_k C^k = 0, \quad \nabla_k \widetilde{C}^k = 0,$$

where the components C^k and $\widetilde{C}^k$ can be found from the relations

$$v_k C^k = \langle Q_{\mathrm{v}} \overset{*}{\mathbf{\Psi}} | \widetilde{Q}_{\mathrm{t}} \mathbf{\Psi} \rangle, \quad v_k \widetilde{C}^k = \langle \widetilde{Q}_{\mathrm{v}} \overset{*}{\mathbf{\Psi}} | Q_{\mathrm{t}} \mathbf{\Psi} \rangle. \tag{31}$$

To find the dual pseudo-charge densities and the dual currents we set

$$C^k = t^k(\mathbf{t}, \mathbf{C}) + C^k - t^k(\mathbf{t}, \mathbf{C}) = \rho t^k + J^k,$$

$$\widetilde{C}^k = t^k(\mathbf{t}, \widetilde{\mathbf{C}}) + \widetilde{C}^k - t^k(\mathbf{t}, \widetilde{\mathbf{C}}) = \widetilde{\rho} t^k + \widetilde{J}^k.$$

Since $\rho = t_k C^k, \quad \widetilde{\rho} = t_k \widetilde{C}^k$, we have from (31)

$$\rho = \langle Q_{\mathrm{t}} \overset{*}{\mathbf{\Psi}} | \widetilde{Q}_{\mathrm{t}} \mathbf{\Psi} \rangle = (Q_{\mathrm{t}} Q_{\mathrm{t}} \overset{*}{\mathbf{\Psi}} | \widetilde{Q}_{\mathrm{t}} \widetilde{Q}_{\mathrm{t}} \mathbf{\Psi}) = (\overset{*}{\mathbf{\Psi}} | \mathbf{\Psi}) = \widetilde{\rho}.$$

We have arrived at the needed fundamental result because $(\overset{*}{\mathbf{\Psi}} | \mathbf{\Psi}) \geq 0$ and $(\overset{*}{\mathbf{\Psi}} | \mathbf{\Psi}) = 0$ provides that $\mathbf{\Psi} = 0$.

The equations of the complex spin field which we connect with the new representations of the bilateral symmetry in the free case look like as follows:

$$(\Pi + \frac{1}{2} \varphi Q_{\mathrm{t}}) \mathbf{\Psi} = m \mathbf{\Psi}, \tag{32}$$

$$(\widetilde{\Pi} + \frac{1}{2} \varphi \widetilde{Q}_{\mathrm{t}}) \mathbf{\Psi} = -m \mathbf{\Psi}, \tag{33}$$

and formally coincide with equations (23) and (24) but in this case we do not need to put an additional condition on the complex spin field $\mathbf{\Psi}$.

Let B_i be components of the potential b of the pseudo-electromagnetic field. In accordance with (31), we write down the dual Lagrangians of interactions of the pseudo-electromagnetic field and the complex spin field:

$$\pounds_{int} = q\, B_i C^i = q \langle Q_{\mathrm{b}} \overset{*}{\mathbf{\Psi}} | \widetilde{Q}_{\mathrm{t}} \mathbf{\Psi} \rangle, \quad \widetilde{\pounds}_{int} = q\, B_i \widetilde{C}^i = \langle \widetilde{Q}_{\mathrm{b}} \overset{*}{\mathbf{\Psi}} | Q_{\mathrm{t}} \mathbf{\Psi} \rangle,$$

where q is a constant of interaction. Taking into account these interactions, we change equations (32) and (33) by the simple substitution $\breve{\nabla}_i \rightarrow \breve{\nabla}_i - iq\, B_i$.

Besides, it is very important to emphasize that the dual Lagrangians of the complex spin field are invariant with respect to the transformations of a wider group of symmetry. Really, we have the commutation relations

$$\Pi \widetilde{Q}_{\mathrm{t}} - \widetilde{Q}_{\mathrm{t}} \Pi = 0, \quad \widetilde{\Pi} Q_{\mathrm{t}} - Q_{\mathrm{t}} \widetilde{\Pi} = 0$$

and

$$\Pi\widetilde{J} - \widetilde{J}\Pi = 0, \quad \widetilde{\Pi}J - J\widetilde{\Pi} = 0.$$

Setting

$$\widetilde{I} = \widetilde{Q}_{\mathbf{t}}, \quad \widetilde{I}\widetilde{J} = -\widetilde{J}\widetilde{I} = \widetilde{K}, \quad I = Q_{\mathbf{t}}, \quad IJ = -JI = K,$$

$$\Lambda = \Lambda_1 E + \Lambda_2 I + \Lambda_3 J + \Lambda_4 K, \quad |\Lambda|^2 = {\Lambda_1}^2 + {\Lambda_2}^2 + {\Lambda_3}^2 + {\Lambda_4}^2,$$

$$\widetilde{\Lambda} = \widetilde{\Lambda}_1 E + \widetilde{\Lambda}_2 \widetilde{I} + \widetilde{\Lambda}_3 \widetilde{J} + \widetilde{\Lambda}_4 \widetilde{K}, \quad |\widetilde{\Lambda}|^2 = \widetilde{\Lambda}_1^2 + \widetilde{\Lambda}_2^2 + \widetilde{\Lambda}_3^2 + \widetilde{\Lambda}_4^2$$

and taking into account that

$$\widetilde{Q}_{\mathbf{t}}\widetilde{J} + \widetilde{J}\widetilde{Q}_{\mathbf{t}} = 0, \quad Q_{\mathbf{t}}J + JQ_{\mathbf{t}} = 0,$$

we discover that the Lagrangian $\mathscr{L}_{\mathbf{t}}$ is invariant with respect to the transformations

$$\mathbf{\Psi}' = \frac{\widetilde{\Lambda}}{|\widetilde{\Lambda}|}\mathbf{\Psi}$$

and the dual Lagrangian $\widetilde{\mathscr{L}}_{\mathbf{t}}$ is invariant with respect to the transformations

$$\mathbf{\Psi}' = \frac{\Lambda}{|\Lambda|}\mathbf{\Psi}.$$

This symmetry opens a new form of interactions which will be carefully considered elsewhere.

To apply formalisms of the vector analysis to equation (32), we set (as in the case of the real spin field)

$$\kappa = t^i\psi_i, \quad \mu = \psi, \quad \lambda = \frac{1}{3!}t_l e^{ijkl}\psi_{ijk}, \quad \nu = \frac{1}{4!}e^{ijkl}\psi_{ijkl},$$

$$\mathrm{K}^i = h^i_m\psi^m, \quad \mathrm{M}^i = t_k\psi^{ki}, \quad \mathrm{L}^i = \frac{1}{3!}h^i_m e^{mjkl}\psi_{jkl}, \quad \mathrm{N}^i = t_k\widetilde{\psi}^{ki},$$

where $h^i_j = \delta^i_j - t^i t_j, \quad \widetilde{\psi}^{ki} = \frac{1}{2}e^{kijl}\psi_{jl}$. The inverse mapping has the form

$$\psi = \mu, \quad \psi_i = \mathrm{K}_i + \kappa t_i, \quad \psi_{ij} = t_i\mathrm{M}_j - t_j\mathrm{M}_i + e_{ijkl}t^k\mathrm{N}^l,$$

$$\psi_{ijk} = e_{mijk}\mathrm{L}^m + t^m e_{ijkm}\lambda, \quad \psi_{ijkl} = e_{ijkl}\,\nu.$$

The equations for the complex spin field read

$$\begin{aligned}(\nabla_{\mathbf{t}} + \tfrac{1}{2}\varphi)\kappa &= \operatorname{div}\mathbf{K} - m\,\mu\\ (\nabla_{\mathbf{t}} + \tfrac{1}{2}\varphi)\lambda &= \operatorname{div}\mathbf{L} - m\,\nu\\ (\nabla_{\mathbf{t}} + \tfrac{1}{2}\varphi)\mu &= \operatorname{div}\mathbf{M} + m\,\kappa\\ (\nabla_{\mathbf{t}} + \tfrac{1}{2}\varphi)\nu &= \operatorname{div}\mathbf{N} + m\,\lambda\end{aligned} \tag{34}$$

$$\begin{aligned}(\nabla_{\mathbf{t}} + \tfrac{1}{2}\varphi)\mathbf{K} &= -\operatorname{rot}\mathbf{L} + \operatorname{grad}\kappa + m\,\mathbf{M}\\ (\nabla_{\mathbf{t}} + \tfrac{1}{2}\varphi)\mathbf{L} &= \operatorname{rot}\mathbf{K} + \operatorname{grad}\lambda + m\,\mathbf{N}\\ (\nabla_{\mathbf{t}} + \tfrac{1}{2}\varphi)\mathbf{M} &= \operatorname{rot}\mathbf{N} + \operatorname{grad}\mu - m\,\mathbf{K}\\ (\nabla_{\mathbf{t}} + \tfrac{1}{2}\varphi)\mathbf{N} &= -\operatorname{rot}\mathbf{M} + \operatorname{grad}\nu - m\,\mathbf{L}.\end{aligned} \tag{35}$$

The dual operators J and $\widetilde{J}$ are tightly connected with the concept of orientation. The existence of the neutrino gives right to call these dual operators by the operators of the neutrino charge. Since $J^2 = \widetilde{J}^2 = -E$, we define the states with the neutrino charge equal to ± 1 by the algebraic equations

$$J\boldsymbol{\Psi} = \pm i\,\boldsymbol{\Psi}, \quad \widetilde{J}\boldsymbol{\Psi} = \pm i\,\boldsymbol{\Psi}.$$

For the state $\widetilde{J}\boldsymbol{\Psi} = i\boldsymbol{\Psi}$ we get

$$\kappa = i\,\lambda, \quad \mu = i\,\nu, \quad \mathrm{K}_i = i\,\mathrm{L}_i, \quad \mathrm{M}_i = i\,\mathrm{N}_i$$

and the equations of the spin field take in this case the form

$$\begin{aligned}(\nabla_{\mathbf{t}} + \tfrac{1}{2}\varphi)\kappa &= \operatorname{div}\mathbf{K} - m\,\mu\\ (\nabla_{\mathbf{t}} + \tfrac{1}{2}\varphi)\mu &= \operatorname{div}\mathbf{M} + m\,\kappa\end{aligned} \tag{36}$$

$$\begin{aligned}(\nabla_{\mathbf{t}} + \tfrac{1}{2}\varphi)\mathbf{K} &= i\operatorname{rot}\mathbf{K} + \operatorname{grad}\kappa + m\,\mathbf{M}\\ (\nabla_{\mathbf{t}} + \tfrac{1}{2}\varphi)\mathbf{M} &= -i\operatorname{rot}\mathbf{M} + \operatorname{grad}\mu - m\,\mathbf{K}.\end{aligned} \tag{37}$$

Taking into account the duality of Time and the bilateral symmetry, we can distinguish four levels of organization of matter at the quantum level.

The real spin field and usual Time. Particles of these fields carry the electric charge and can explain the so called “pairing of the electrons” in the physics of atoms and molecules.

The real spin field and dual Time. This is an unusual situation and a question of its physical interpretation is under consideration.

The complex spin field and usual Time. Particles of this field carry the pseudo-charge, the electric charge and the neutrino charge. This level of organization corresponds to the electroweak sector of the Standard Model.

The complex spin field and dual Time. Particles of this field carry the pseudo-charge, the electric charge and the neutrino charge. This level of organization corresponds to the physics of so called strong interactions.

Dual Time naturally explains the confinement and the baryon number conservation. Indeed, we cannot use the concept of force to explain the confinement because for any force there is a more powerful one. But the dual Time in a bundle with the compact physical space do this naturally (the connection of dual Time and rotational motion). The baryon number conservation simply expresses in a symbolic form the existence of the second Time in the nature and nothing else. The fundamental quark-lepton symmetry means that leptons live in the usual Time and quarks live in the area of dual Time.

The algebra of the electric charge and the neutrino charge operators $Q_e = \widetilde{Q}_{\mathrm{t}}$ and $Q_\nu = \widetilde{J}$ is very simple

$$Q_e{}^2 = -E, \quad Q_\nu{}^2 = -E, \quad Q_e Q_\nu + Q_\nu Q_e = 0.$$

With this algebra one can easy derive that the observed generations of the quarks and leptons of the Standard Model can be considered as different states of the complex spin field and the number of these families equals four. Indeed, let us introduce the projection operators:

$$P_+ = \frac{1}{2}(E + iQ_e), \quad P_- = \frac{1}{2}(E - iQ_e),$$

$$R_+ = \frac{1}{2}(E + iQ_\nu), \quad R_- = \frac{1}{2}(E - iQ_\nu).$$

For the identity operator we have the following expansions:

$$E = P_+ + P_- = P_+R_+ + P_+R_- + P_-R_+ + P_-R_-$$

and

$$E = R_+ + R_- = R_+P_+ + R_+P_- + R_-P_+ + R_-P_-.$$

The algebra in question also explains the so called "pairing" of the electron and neutrino of the Standard Model.

FAMILIAR SPACE-TIME STRUCTURE

All our previous considerations were performed under very general conditions, when metric field $g_{ij}(x)$ and gradient of Time constrained by the only condition $g^{ij}\partial_i f \partial_j f = 1$. Now it is time to present equations of the spin field under the conditions:

$$g_{ij} = \delta_{ij}, \quad f(x) = a_i x^i, \quad f(x) = \sqrt{(x^1)^2 + (x^2)^2 + (x^3)^2 + (x^4)^2}.$$

Below we exhibit the wave equations of quantum mechanics of particles with the Emergent Spin one half with respect to the Time $t = a_i x^i$:

$$\begin{aligned} D\,\kappa &= \operatorname{div}\mathbf{K} - m\,\mu \\ D\,\lambda &= \operatorname{div}\mathbf{L} - m\,\nu \\ D\,\mu &= \operatorname{div}\mathbf{M} + m\,\kappa \\ D\,\nu &= \operatorname{div}\mathbf{N} + m\,\lambda \end{aligned} \tag{38}$$

$$\begin{aligned} D\,\mathbf{K} &= -\operatorname{rot}\mathbf{L} + \operatorname{grad}\kappa + m\,\mathbf{M} \\ D\,\mathbf{L} &= \operatorname{rot}\mathbf{K} + \operatorname{grad}\lambda + m\,\mathbf{N} \\ D\,\mathbf{M} &= \operatorname{rot}\mathbf{N} + \operatorname{grad}\mu - m\,\mathbf{K} \\ D\,\mathbf{N} &= -\operatorname{rot}\mathbf{M} + \operatorname{grad}\nu - m\,\mathbf{L}, \end{aligned} \tag{39}$$

where $D = a^i\partial_i$. Setting $a_i = (0, 0, 0, 1)$ and hence, $t = x^4$ we write down eqs. (38) and (39) in the simplest form

$$\begin{aligned} \frac{\partial\kappa}{\partial t} &= \operatorname{div}\mathbf{K} - m\,\mu \\ \frac{\partial\lambda}{\partial t} &= \operatorname{div}\mathbf{L} - m\,\nu \\ \frac{\partial\mu}{\partial t} &= \operatorname{div}\mathbf{M} + m\,\kappa \\ \frac{\partial\nu}{\partial t} &= \operatorname{div}\mathbf{N} + m\,\lambda \end{aligned} \tag{40}$$

$$\begin{aligned} \frac{\partial\mathbf{K}}{\partial t} &= -\operatorname{rot}\mathbf{L} + \operatorname{grad}\kappa + m\,\mathbf{M} \\ \frac{\partial\mathbf{L}}{\partial t} &= \operatorname{rot}\mathbf{K} + \operatorname{grad}\lambda + m\,\mathbf{N} \\ \frac{\partial\mathbf{M}}{\partial t} &= \operatorname{rot}\mathbf{N} + \operatorname{grad}\mu - m\,\mathbf{K} \\ \frac{\partial\mathbf{N}}{\partial t} &= -\operatorname{rot}\mathbf{M} + \operatorname{grad}\nu - m\,\mathbf{L} \end{aligned} \tag{41}$$

.

Let A_i be the vector potential of the electromagnetic field. The gauge invariant tensor of the electromagnetic field is defined as usual $F_{ij} = \partial_i A_j - \partial_j A_i$.

The strength of the electric field is a general covariant and gauge invariant quantity that is defined by the equation $E_i = a^k F_{ik}$.

The rotor of the vector field $\mathbf{A} = (A_1, A_2, A_3, A_4)$ is defined as a vector product of ∇_4 and $\mathbf{A}$

$$\mathrm{rot}\,\mathbf{A} = \nabla_4 \times \mathbf{A}, \quad (\mathrm{rot}\,\mathbf{A})^i = e^{ijkl} a_j \partial_k A_l = \frac{1}{2} e^{ijkl} a_j (\partial_k A_l - \partial_l A_k),$$

where e^{ijkl} are the contravariant components of the Levi-Civita tensor normalized as $e_{1234} = 1$. The general covariant and gauge invariant definition of the magnetic field strength is given by the formula $\mathbf{H} = \mathrm{rot}\,\mathbf{A}, \quad H^i = (\mathrm{rot}\,\mathbf{A})^i$. Thus, $H_i = a^k \overset{*}{F}_{ik}$, where $\overset{*}{F}_{ij} = g_{ik} g_{jl} \overset{*}{F}{}^{kl} = \frac{1}{2} g_{ik} g_{jl} e^{klmn} F_{mn}$. It is evident that vectors $\mathbf{E}$ and $\mathbf{H}$ are orthogonal to $\mathbf{a}$

$$(\mathbf{a}|\mathbf{E}) = 0, \quad (\mathbf{a}|\mathbf{H}) = 0.$$

The Maxwell equations read:

$$D\,\mathbf{H} = -\mathrm{rot}\,\mathbf{E}, \tag{42}$$

$$D\,\mathbf{E} = \mathrm{rot}\,\mathbf{H}, \tag{43}$$

$$(\nabla_4|\mathbf{E}) = 0, \quad (\nabla_4|\mathbf{H}) = 0. \tag{44}$$

Now it is important to show the definition of the interval in the space in question. The interval in R^4 is defined as follows. Let

$$\mathbf{x}_s = 2(\mathbf{a}|\mathbf{x})\,\mathbf{a} - \mathbf{x}$$

be the vector symmetrical to the vector $\mathbf{x}$ with respect to the vector $\mathbf{a}$. Then in the coordinates x^1, x^2, x^3, x^4 the interval can be written as follows:

$$s^2 = (\mathbf{x}|\mathbf{x}_s) = 2(\mathbf{a}|\mathbf{x})^2 - (\mathbf{x}|\mathbf{x}) = (\mathbf{x}|\mathbf{x}) \cos 2\theta,$$

where θ is an angle between $\mathbf{a}$ and $\mathbf{x}$. It is easy to see that in the system of coordinates x, y, z, t, where $\mathbf{a} = (0, 0, 0, 1)$

$$s^2 = t^2 - x^2 - y^2 - z^2.$$

We see that the first space-time structure and the bilateral symmetry presuppose the existence of the familiar space-time structure. The special theory of relativity is discovered here as a spontaneous breaking of isotropy of the four-dimensional Euclidian space. The bilateral symmetry defines the auxiliary metric in this case as usual

$$\overline{g}_{ij} = 2a_i a_j - \delta_{ij}.$$

Thus, when we put in correspondence to any point of R^4 a moment of time t by the equation

$$t = a_i x^i,$$

we uncover the underlying structure of the special theory of relativity. We can take another gradient of Time b_i which is different from a_i and consider the equations in question associated with b_i. One can show that these equations are equivalent since there is one-to-one and smooth transformation of R^4 onto itself which translates b_i into a_i.

Fundamental Spin

We have demonstrated that in a certain sense the well-known idea of "rotating rigid body" (also mentioned as the Top) of classical mechanics is as fundamental as the idea of "mass point", i.e., the first concept can be reduced to the second one at the fundamental (field -theoretical) level and dual Time plays a key role in this consideration. The duality of Time was presented above and here we exhibit the dynamical equations of the complex spin field which describe the phenomenon of Fundamental Spin:

$$\begin{aligned} \tfrac{1}{\tau}(D+\tfrac{3}{2})\kappa &= \operatorname{div}\mathbf{K} - m\,\mu \\ \tfrac{1}{\tau}(D+\tfrac{3}{2})\lambda &= \operatorname{div}\mathbf{L} - m\,\nu \\ \tfrac{1}{\tau}(D+\tfrac{3}{2})\mu &= \operatorname{div}\mathbf{M} + m\,\kappa \\ \tfrac{1}{\tau}(D+\tfrac{3}{2})\nu &= \operatorname{div}\mathbf{N} + m\,\lambda \end{aligned} \tag{45}$$

$$\begin{aligned} \tfrac{1}{\tau}(D+\tfrac{3}{2})\mathbf{K} &= -\operatorname{rot}\mathbf{L} + \operatorname{grad}\kappa + m\,\mathbf{M} \\ \tfrac{1}{\tau}(D+\tfrac{3}{2})\mathbf{L} &= \operatorname{rot}\mathbf{K} + \operatorname{grad}\lambda + m\,\mathbf{N} \\ \tfrac{1}{\tau}(D+\tfrac{3}{2})\mathbf{M} &= \operatorname{rot}\mathbf{N} + \operatorname{grad}\mu - m\,\mathbf{K} \\ \tfrac{1}{\tau}(D+\tfrac{3}{2})\mathbf{N} &= -\operatorname{rot}\mathbf{M} + \operatorname{grad}\nu - m\,\mathbf{L}, \end{aligned} \tag{46}$$

where $\tau = \sqrt{(x^1)^2 + (x^2)^2 + (x^3)^2 + (x^4)^2}$ is the dual Time and $D = x^i \partial_i$. The other operators are defined as follows:

$$(\mathrm{rot}\,\mathbf{M})^i = \frac{1}{\tau} e^{ijkl} x_j \partial_k M_l, \quad (i,\, j,\, k,\, l = 1, 2, 3, 4)$$

where e^{ijkl}, is the Levi-Civita tensor normalized as follows $e_{1234} = \sqrt{g} = 1$;

$$(\mathrm{grad}\varphi)_i = \triangle_i\, \varphi, \quad \triangle_i = \partial_i - \frac{x_i}{\tau^2} D, \quad \mathrm{rot}\,\mathrm{grad} = 0, \quad \mathrm{div}\,\mathrm{rot} = 0$$

and $(\mathbf{x}|\mathbf{K}) = (\mathbf{x}|\mathbf{L}) = (\mathbf{x}|\mathbf{M}) = (\mathbf{x}|\mathbf{N}) = 0$. The Maxwell equations for dual photons read

$$\frac{1}{\tau}(D+2)\mathbf{H} = -\mathrm{rot}\,\mathbf{E}, \quad \frac{1}{\tau}(D+2)\mathbf{E} = \mathrm{rot}\,\mathbf{H}, \tag{47}$$

$$(\mathbf{x}|\mathbf{E}) = (\mathbf{x}|\mathbf{H}) = 0, \quad \mathrm{div}\,\mathbf{E} = \mathrm{div}\,\mathbf{H} = 0, \tag{48}$$

where

$$E_i = \frac{x^k}{\tau} F_{ik}, \quad H_i = \frac{x^k}{\tau} \overset{*}{F}_{ik}, \quad F_{ij} = \partial_i A_j - \partial_j A_i, \quad \overset{*}{F}_{ij} = \frac{1}{2} e_{ijkl} F^{kl}.$$

It is natural to put forward the conjecture that equations (45),(46) and (47),(48) are a key to discover the internal structure of hadrons and nuclei.

CONCLUSION

Since searches for physics beyond the Standard Model have not been successful so far, it makes sense to consider new approaches to fundamental physics as a whole. Spin is already a great unifying principle in theoretical physics, but its potential is far from being exhausted since the concept of Emergent Spin, as it is demonstrated above, provides new concepts and equations for the physics of elementary particles. Equations of the Fundamental Spin are the result of a shift from the concepts and equations related with the idea of "rotating rigid body" of classical mechanics to the corresponding natural concepts and equations at the fundamental (field-theoretical) level.

At the present time, quantum chromodynamics has no alternative, but in the framework of this theory we have no answer to the set of principle questions.

To explain the confinement mechanism is a very important, but still unresolved problem, in particle physics. Hence, new approaches and concepts are desirable in any case. From this point of view our suggestion to consider leptons on the ground of the one space-time structure and connect quarks (dual particles) with the dual space-time structure on the same four-dimension Euclidian space appears to be quite timely. We put forward a conjecture that this beautiful duality is adequate to the nature of things. The problem of time and everything connected with this topic are up to now significant. The results obtained are significant because they give a simple and evident explanation of lepton-quark symmetry, quark confinement and baryon number conservation. The confinement and the baryon number conservation simply mean that the quarks (dual particles) cannot change their space-time structure because they are doomed for the eternal natural rotation. Finally, we stress that the dual space-time structure is an evident guide in the attempts to find Lorentz violation (meaning a research program on this topic proposed by Alan Kostelesky).

REFERENCES

[1] Uhlenbeck G. E., and Goudsmit, S. 1926. “Spinning Electrons and the Structure of Spectra.” *Nature* 117(2938): 264-65.

[2] Pestov, Ivanhoe. 2005. “Field Theory and the Essence of Time.” In *Horizons in World Physics, Volume 248. Spacetime Physics Research Trends,* edited by Albert Reimer, 1-29. New York: Nova Science Publishers, Inc.

[3] DeWitt Bryce S. 1965. *Dynamical Theory of Groups and Fields.* Gordon and Breach New York.

[4] Pestov, Ivanhoe. 2006. “Dark Matter and Potential Fields.” In *Dark Matter: New Research,* edited by J. Val Blain, 73-90. New York: Nova Science Publishers, Inc.

[5] Biedenharn, L.C., and Louck, J.D. 1981. *Angular Momentum in Quantum Physics.* Massachusetts: Addison-Wesley Publishing Company Reading.

In: Horizons in World Physics. Volume 302 ISBN: 978-1-53617-180-8
Editor: Albert Reimer

Chapter 5

PROPERTIES OF ATOMIC NUCLEI ENSEMBLES (MOLECULAR CLUSTERS AND DOMAINS) AS A NEW FORM OF MATTER

Kristina Zubow[1,*], Anatolij Zubow[2] and Viktor Anatolievich Zubow[1]
[1]R&D, Zubow Consulting, Germany
[2]Department of Telecommunication Networks Group, TU Berlin, Berlin, Germany

ABSTRACT

The method of gravitational mass spectroscopy (GMS) was used to study the effect of weak shock waves (SW) on the formation of atomic nuclei concentration (ANC) in water, ethanoic acid, acrylamide, carbonic acid and protein (gelatin). The phenomenon of energy storage inside the ANC ensembles was discovered, and after the action of SW was switched off, a sharp release of energy in the form of a strong secondary SW

* Corresponding Author's Email: Heide-Lore@zubow.de.

"running through" all ANC in the ensemble and activation/destruction of the most unstable mass cluster. The selectivity of the process was achieved by selecting the intensity of the SW, as well as their shape and time of influence. The structures of some experimentally discovered clusters were presented.

Keywords: molecular clusters and domains, ensembles, acetic acid, acrylamide, water, gelatin, carbonic acid, shock waves

1. Introduction

Earlier, we found that ensembles of nuclear clots in molecular matter (liquids, polymers) form stable formations with individually expressed properties. This concerned energy and its distribution between individual sub-ensembles and clusters/domains, remote interaction of ensemble elements with analogs of the environment, A fundamental possibility was found to influence the entire ensemble of mass clots or its components (clusters and their sub ensembles) [1, 2]. Of practical interest is the use of this phenomenon for the forced activation of individual domains in proteins and clusters in liquids. In [2], we showed the effect of weak sinusoidal SW on the stabilization of a particular morphic structure of water. However, the most interesting prospects are revealed by the impact on the polymorphic structures of specific forms of SW.

The aim of this work was to search for the soft physical effects of various forms of SW on molecular ensembles in liquids and polymers for the selective control of the energy state of its individual structural components.

2. Materials and Method

The following were selected as objects of study: discilled water aged for 24 months (stored in glass vessel all in grounded iron box) (Z = 6.4E-15 N/m), acetic acid (aged for 6 months) (Z = 5.9E-15 N/m), acrylamide, crystalline (p. A, Z = 6.9E-15 N/m), carbonic acid solution aged over 12 months (Z = 6E-15 N/m) and gelatin (transparent film, Dr. Oetker, Germany, Z = 6.55E-15 N/m). Figure 1 shows an installation diagram for studying the influence of SW on these objects. The distance beween the sound frequency generator (20 ... 22 kHz) and solution surface was 3…8 mm. Miniature sound generators (0.5 W) were used from a personal computer. The notebook was powered by a laptop. A 20 kHz frequency was chosen as an example only. To calculate the masses of ANC, the first Zubow equation was used [1]. The Δf value was obtained by subtracting the f value for the clusters before and after irradiation. Negative Δf - values relate to collapsed (dense) clusters, and positive values to expand (loose). Cluster masses (*m*) are given in Daltons. The technique and essence of the GMS method were described in the monograph [1]. The object of study was protected from external energy physical influences (electromagnetic radiation, mechanical fields and strong pressure drops). Recall that the ANC in baryonic matter (molecular clusters, domains in polymers), like galaxies clusters, were formed under the influence of gravitational noise (GN) of the Universe. That is, the GN of the Universe forbid and allow the formation of certain ANC. Thus, the existence of clusters and domains in molecular matter is strictly determined. Their oscillation frequencies and masses are described by the first and second Zubow equations [1].

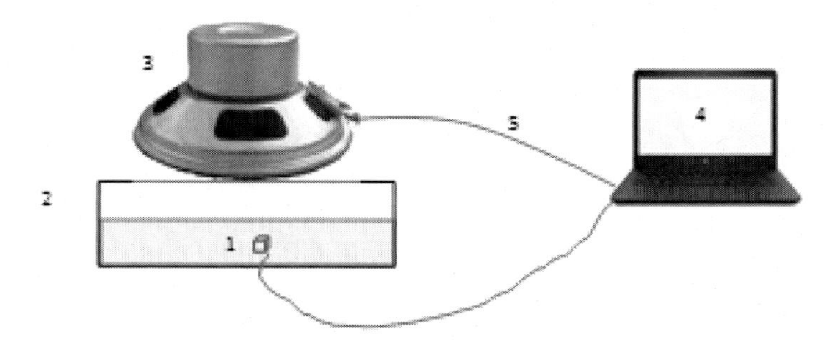

Figure 1. Scheme of the effects of SW on samples. 1 - liquid (sample) and GMS sensor, 2 - vessel, 3 - 0.5 W (speaker), 4 - PC, 5 - interface. 20 kHz.

The number of kinds of signals in the spectrum of the GMS (N) represents a family of clusters and constellations of domains in the material. The fraction of dense domains (D_c, %) was calculated as the ratio of the number of dense domains (*f*) to their total number in the studied ensemble (N). The average mass of domains in the ensemble (M_{GMS}, Dalton) is the sum of the product of shares of domains (abs *f*) by their masses (*m*, Da). Z is the Zubow force constant [1].

3. Results and Its Discussion

Figure 2 shows the GMS spectra of water exposed to various forms of SW with a frequency of 20 kHz. It can be seen that already a short-term impact of SW of various shapes has a different effect on the long-range order (LRO) of water. If in the initial sample there was a clear equilibrium between the dense and loose structures in the base cluster of 11 ± 1 molecules [3], then the action of the shock wave violated it and new signals from the dominant collapsed and expanded structures appeared. All types of SW transfer equilibrium for clusters of 25 molecules into the mode of dominance of loose structures (*f*). The dominance of loose structures in the region of small clusters appears, while in the region of giant sub micellar, micellar, and super micellar structures (log $m > 5$), new ANC with different dominance of dense or loose structures appear. Thus, giant micelle structures of 2,170,924 and 19,538,318 molecules compacted (-*f*). An increase in the time of exposure of these forms of SW to water to 10 min practically does not lead to changes in the LRO, but only under the soft effects of weak shock waves. Selective activation of individual ANC and change their shape (expanded-collapsed), in our opinion, can exert a strong influence on all physical and chemical processes involving water. It should be borne in mind that a decrease in SW power and an increase in the time of their exposure enhances structural changes in the LRO of water.

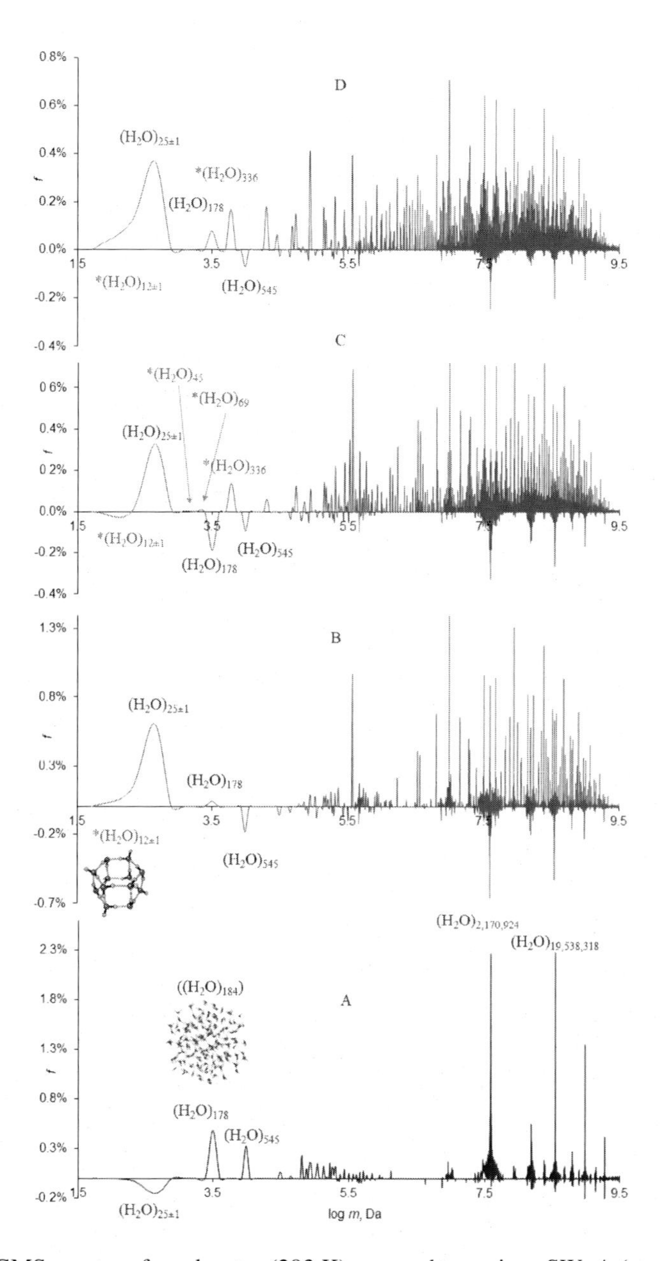

Figure 2. GMS spectra of aged water (293 K) exposed to various SW. A (start): N = 2408, M_{GMS} = 512,578,961, Da, D_c = 10%; B (1 min, 20 kHz, pulse): N = 3199, M_{GMS} = 424,388,232, Da, D_c = 24%; C (1 min, 20 kHz, sawtooth): N = 3199, M_{GMS} = 382,635,145, Da, D_c = 20%; D (1 min, 20 kHz, square): N = 3359, M_{GMS} = 394,267,969, Da, D_c = 12%. The asterisk marks the appearance of new clusters. Water cluster models were kindly provided by professors Lenz and Chaplin [4, 5].

The problems of stabilization of the LRO of chemical compounds dissolved in water are somewhat more complicated. For example, acrylamide and ethanoic acid. Here, one has to subtract the GMS spectra of the solvent from the GMS spectra of the solution under adequate experimental conditions. Thus, the dissolution of crystalline acrylamide in aged water leads to the first unstable thermodynamic state of the liquid. Significant times are required to establish a new thermodynamic equilibrium, which is practically not interesting. Therefore, we subtracted the spectra of liquids for the same conditions. For example, from the GMS spectrum of a solution of acrylamide subjected to a sawtooth SW (f_s) for 3 minutes, the signals of the GMS spectrum of aged water, subjected to a 3-minute exposure to sawtooth SW under the same external energy influences, (f_w) were subtracted. The difference in the values of $f_s - f_w = \Delta f$ was used to interpret the LRO spectra of the acrylamide molecules in the solution. With all the criticism of such a procedure, it turned out to be practically acceptable [1, 2]. Table 1 shows the integrated characteristics of the ensemble of all clusters in water and an aqueous solution of acrylamide before and after exposure to it by sinusoidal SW with a frequency of 20 kHz.

As follows from Table 1, the influence of soft SW sinusoidal shape strongly affects the LRO of the solution. The number of cluster types increases sharply, but their average molecular mass, on the contrary, decreases. On the other hand, these processes cause an almost twofold increase in the ensemble of dense structures. It is noteworthy that after disabling the SW, a noticeable post-effect was observed. It was expressed in a slow decrease in the fraction of dense structures. But this process was not monotonous, an increase of D_c by 2% by the 3rd minute indicated the appearance of a strong secondary SW already inside the ensemble due to the system moving to a new thermodynamic equilibrium state. Such strong secondary super SW distribute the energy of the ensemble between its constituent clusters, and an excess of energy destroys the weakest cluster that was not beneficial to the ensemble. We used the post-effect for targeted, selective destruction of one carbonic acid cluster [6], see also below.

Table 1. Integral characteristics of the LRO solution of acrylamide in water before and after exposure to sinusoidal SW (the form is given on the right, for understanding)

	N	M_{GMS}, Da	D_c, %
H_2O	2344	468291117	8
10% solution	2386	732139859	52
1 min 20 kHz	2605	365694549	99 
5 min 20 kHz	2615	365749030	99
1 min, after	2499	491330950	89
3 min, after	2472	495930136	91

It was of interest to understand the role of acrylamide clusters in solution when exposed to various forms of SW. Figure 3 shows such subtraction spectra for a diluted solution of acrylamide in water.

As can be seen, in the initial freshly prepared solution in the region of small clusters there are no signals from the dominance of both loose and dense structures, that is, there is an equilibrium between them, but when exposed of SW, they are violated and corresponding bands appear in the spectra. Pulse and square shock waves activate acrylamide trimers [7]. If pulsed SW led to the dominance of dense, then square ones to loose trimers. In this case, the effect is especially pronounced. However, acrylamide heptamers behave the same way, with the exception of option C (sawtooth SW), these waves collapsed heptamers. All types of SW affect the collapsed-expanded equilibrium for clusters of 12, 19 and 49 acrylamide molecules. The prolonged exposure to shock waves, for example, of a pulsed nature, up to 3 minutes did not significantly affect the nature of the spectra and the main parameters of the ensemble (N, M_{GMS}, and D_c). Of particular note is the effect of SW on the micellar structures of this substance. The harmony that existed from the beginning is completely destroyed. It is difficult to detect any regularities here due to the long relaxation times of polymer systems. However, the termination of the impact of SW slowly returns these structures to their original state. Thus, the choice of SW forms makes it possible to carry out a targeted effect both on the entire ensemble of acrylamide domains, and on its small clusters, their conformation.

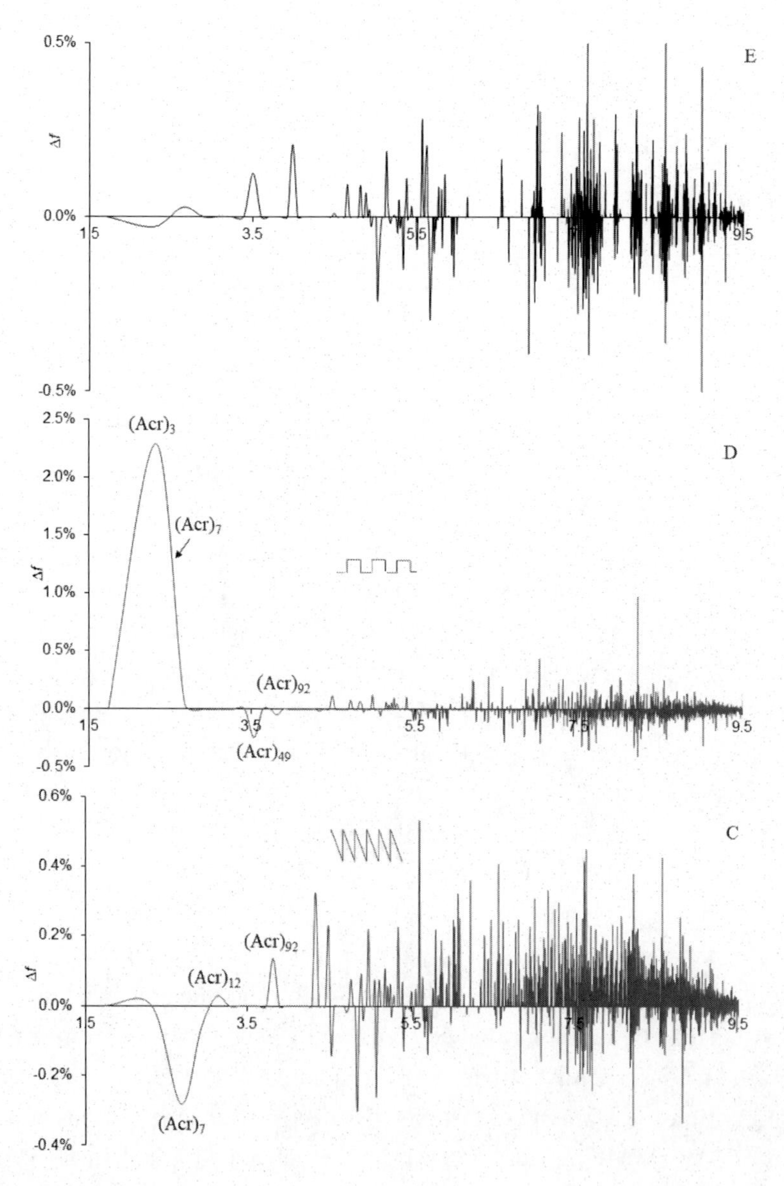

Figure 3. (Continued)

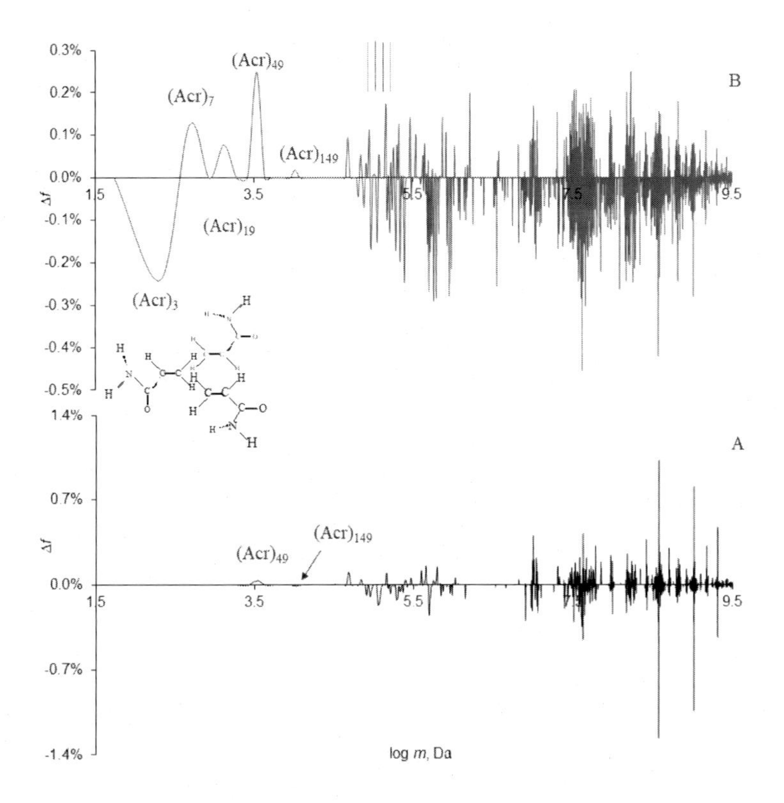

Figure 3. GMS subtraction spectra: A - LRO of acrylamide in its 10% aqueous solution. B - same as A, but subjected to a pulsed SW. C - same as A, but subjected to sawtooth SW, D - same as A, but subjected to a square SW and E - same as A, but after the end of the impact of SW (1 minute later). A: N = 2494, M_{GMS} = 634036615, Da, D_c = 47%; B: N = 2643, M_{GMS} = 459151072, Da, D_c = 65%; C: N = 2253, M_{GMS} = 526704113, Da, D_c = 25%; D: N = 2398, M_{GMS} = 557367665, Da, D_c = 59%; E: N = 2519, M_{GMS} = 550826338, Da, D_c = 46%. B, C, D - everywhere the exposure time was 1 min, 20 kHz. The forms of SW were given on the right for understanding.

For a solution of acrylamide, the effect was expressed in a slight increase in the viscosity of the liquid, apparently, as a result of oligomerization processes. Acrylamide trimer models are shown in Figure 4. An expanded cluster is formed as a result of strong interaction of polar groups with water through hydrogen bonds (Figure 3, D). In a dense cluster, on the contrary, the polar groups of molecules are reoriented to the center, while nonpolar groups interact with water through weak adhesive bonds with hydrogen atoms (Figure 3, C).

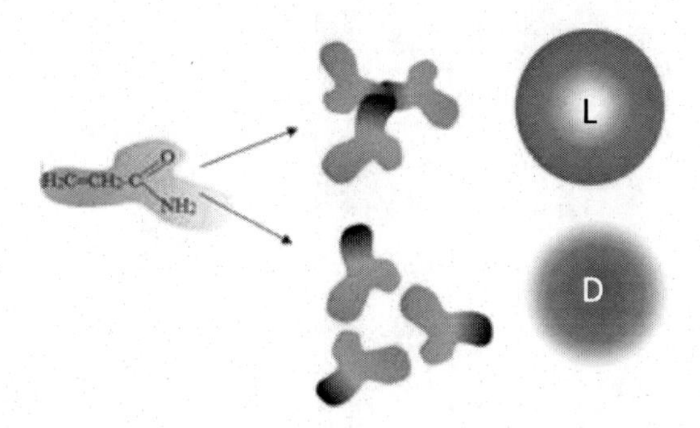

Figure 4. A model of the formation of loose (L) and dense (D) trimers of acrylamide as independent oscillators in water. In the L model, interactions of the cluster with the environment over the intracluster dominate, in the D model, on the contrary.

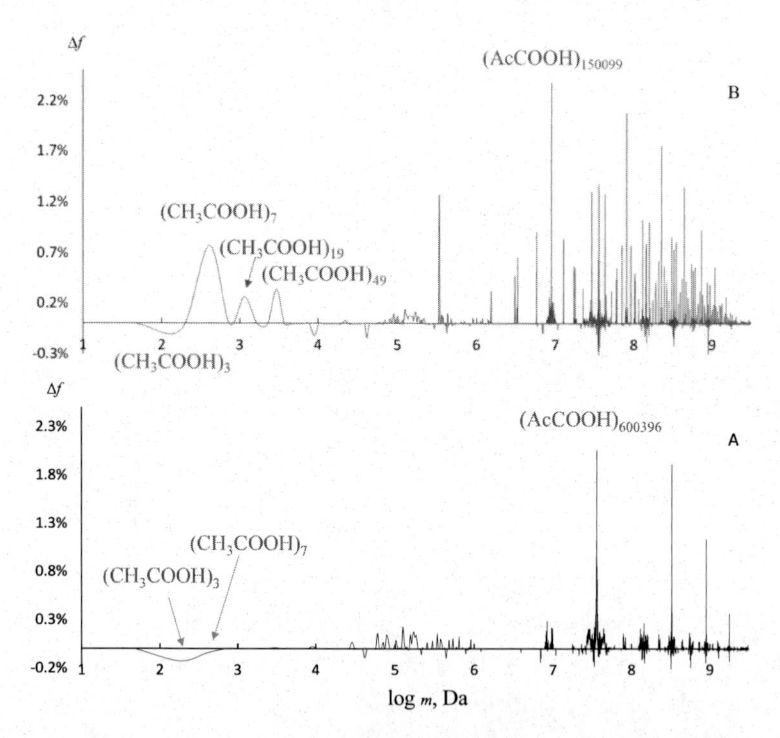

Figure 5. GMS spectra of subtractions. Ethanoic acid (in aqueous 25 wt. % solution, 293 K). A (start): N = 2458, M_{GMS} = 447,083,808, Da, D_c = 14%; B (1 min, 20 kHz, pulse): N = 2478, M_{GMS} = 392,196,226, Da, D_c = 12%.

Figure 5 shows the GMS subtraction spectra for acetic acid clusters in an aged 25% solution. It is seen that the effect of pulsed shock waves causes noticeable changes in the LRO. In the region of low masses, 2 new clusters appear containing 19 and 49 CH_3COOH molecules, respectively, and a cluster of 7 molecules was transferred into a loose state, while the base cluster of 3 molecules [8] has retained its conformation in dense form (Figure 6). In the field of micellar structures, the changes relate to a certain increase in the number of cluster types and a noticeable decrease in their average mass.

Figure 6 shows a model of a basic cluster of acetic acid in water. This is a dense cluster in which the internal interactions of carboxyl groups dominate the interactions of the cluster with their environment. SW in the pulsed forms, unable to loosen this cluster.

Consider the effect of SW forms on a protein solid film, Figure 7.

Their Figure 7 shows that SW have a noticeable effect on the LRO of the protein, but it does not monotonously approach the initial thermodynamic state after the termination of exposure, due to the duration of relaxation processes in the polymer. SW activate the appearance of the base 1RASC domains in expanded form and loosen the next 3RASC domain. In the field of micellar structures, SW cause the appearance of a stable, new sub ensembles of ANC with the dominance of loose structures, Table 2. Table 2 summarizes the main characteristics of the ensemble of domains in protein LRO, Figure 7.

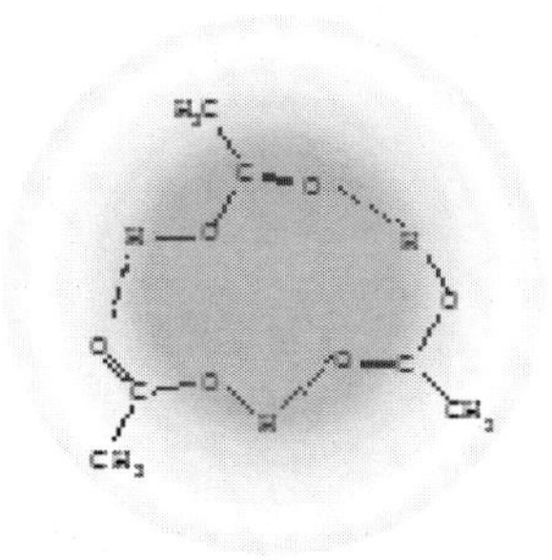

Figure 6. Model of ethanoic acid basic cluster of 3 molecules in collapsed conformation in water.

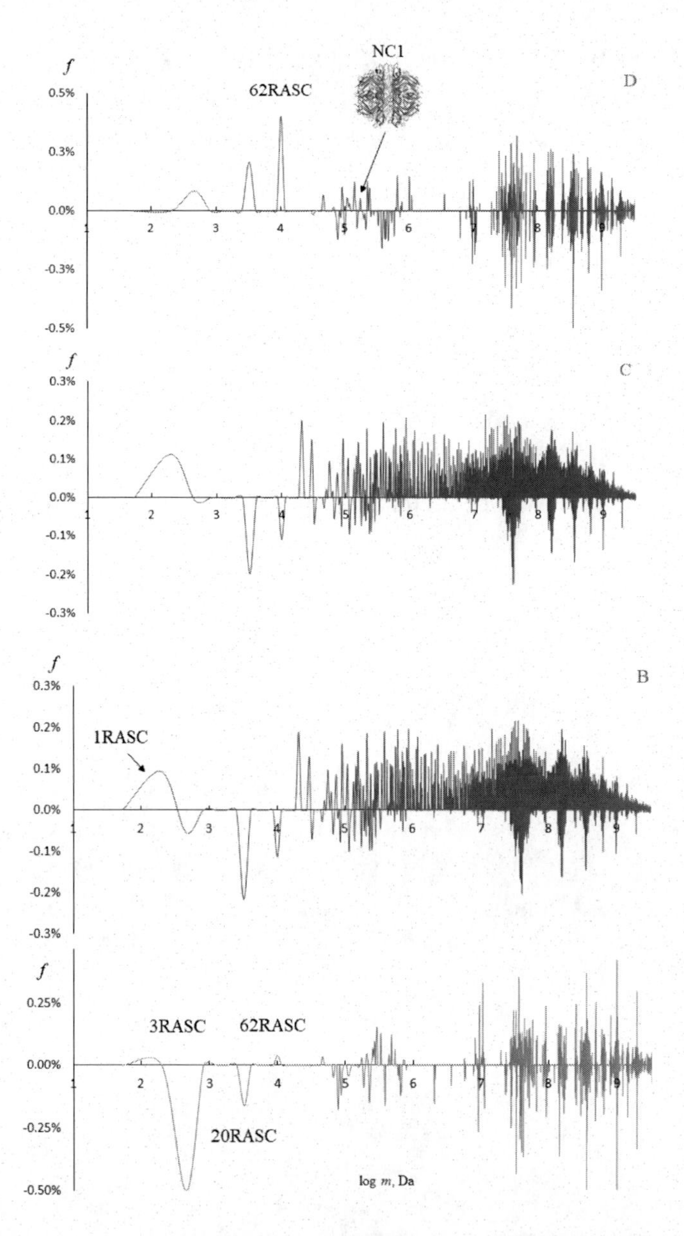

Figure 7. The effect of sinusoidal SW on the LRO in gelatin. The RASC (repeating average structures in collagen chains) consist: Gly 22% - Pro13% - Hyp10% - Glu 9.8% - Ala 8% - Arg 7.6%), 162 Da. A is the original sample; B - A subjected to 1 min irradiation of 20 kHz, C - also, but 10 min. 20 kHz; D - sample C after 120 minutes after turning off the radiation. The structure of the NC1 domain was taken from [9], 172,890 Daltons (Da).

Table 2. The main characteristics of the ensemble of domains in the protein LRO during its processing by sinusoidal SW at 20 kHz, Figure 7

	N	M_{GMS}, Da	D_c, %	*f*, NC1, %
A, original	2438	695337097	47	0.03
1 min	3395	373296012	23	-0.09
3 min	3419	381037497	23	-0.06
5 min	3411	380674826	23	-0.06
10 min	3428	376631625	24	-0.07
1min, after	2462	555358502	50	-0.07
3 min, after	2592	527909939	42	0.00
120 min, after	2355	632962860	54	0.05

Table 3. The main characteristics of the ensemble of domains in protein LRO in the process of interactions it with square SW at 20 kHz

	N	M_{GMS}, Da	D_c, %	*f*, NC1,%	*f*, 1RASC, %	*f*, 3RASC, %
A, original	2438	695337097	47	0.03	0.00	-0.50
1 min	3300	401789596	19	-0.02	0.28	0.29
3 min	3291	405595102	19	-0.02	0.25	0.32
5 min	3279	405920643	19	-0.02	0.27	0.26
10 min	3294	405574940	19	-0.02	0.30	0.32
1 min, after	2466	569661012	38	-0.04	0.00	0.59
3 min, after	2511	569484661	44	-0.13	0.00	0.47
120 min, after	2355	632962860	54	0.05	0.00	0.47

From Table 2 it is seen that the influence of shock waves of a sinusoidal nature sharply increases the number of active types of domains, but decreases their average molecular mass and the part of dense, potential energy rich domains. The cessation of the impact of SW quickly returns the number of domain types to the initial state, but other processes proceed more slowly and not monotonously, for example, conformational changes of NC1 hexamer. Noteworthy is the surge in D_c values by 120 minutes of rest. It may indicate the presence of a secondary SW within the ensemble of domains.

Table 3 summarizes the main characteristics of the ensemble of domains in protein LRO during its processing by square SW.

The effect of square SW also leads to a sharp increase in the number of kinds of active domains, but reduces the molecular mass of domains in the ensemble and reduces the part of energy-rich domains by more than 2 times. Hexamer NC1 compacts, the small domain 1RASC, loosens and quickly recovers in a dense-loose equilibrium. The domain of 3 RASC, on the contrary, quickly loosens and even after 2 hours after that was not able to recover.

Table 4. The main characteristics of the ensemble of domains in protein during its processing by sawtooth SW at 20 kHz

	N	M_{GMS}, Da	D_c,%	*f*, NC1, %	*f*, 1 RASC, %	*f*, 3 RASC, %
A, original	2438	695337097	47	0.03	0.00	-0.50
1 min	2998	440250604	29	-0.06	0.00	0.48
3 min	3012	429258793	29	-0.06	0.00	0.45
5 min	2966	439860693	28	-0.06	0.00	0.51
10 min	3018	426954186	29	-0.06	0.00	0.49
1 min, after	2520	518797589	40	-0.03	0.00	0.61
3 min, after	2520	534767193	40	-0.01	0.00	0.69
120 min, after	2355	632962860	54	0.05	0.00	0.09

Table 5. The main characteristics of the ensemble of domains in the protein during its processing by pulsed SW at 20 kHz

	N	M_{GMS}, Da	D_c, %	*f*, NC1, %	*f*, 1RASC, %	*f*, 3 RASC, %
A, original	2438	695337097	47	0.03	0.00	-0.50
1 min	2675	516072483	35	-0.03	0.00	0.98
3 min	2742	526893362	39	-0.03	0.00	0.91
5 min	2714	517383773	34	-0.02	0.00	0.99
10 min	2694	540649659	34	-0.02	0.00	0.91
1 min, after	2605	499240344	19	0.07	0.00	0.43
3 min, after	2586	483539240	19	0.07	0.00	0.51
120 min, after	2355	632962860	54	0.05	0.00	0.09

Table 4 summarizes the main characteristics of the ensemble of domains in protein during its processing by sawtooth SW.

Sawtooth SW have a much weaker effect on the number of domain types in the ensemble; they do not significantly reduce the molecular mass of domains and the part of their dense structures. In this case, the hexamer was compacted, and the effect on 1 RASC was generally absent. On the other hand, 3RASC was very, loosened and slowly restored even by 2 h after exposure to SW. In general, the influence of sawtooths SW on the entire ensemble can be called soft.

Table 5 summarizes the main characteristics of the ensemble of domains in protein during its processing by pulsed SW.

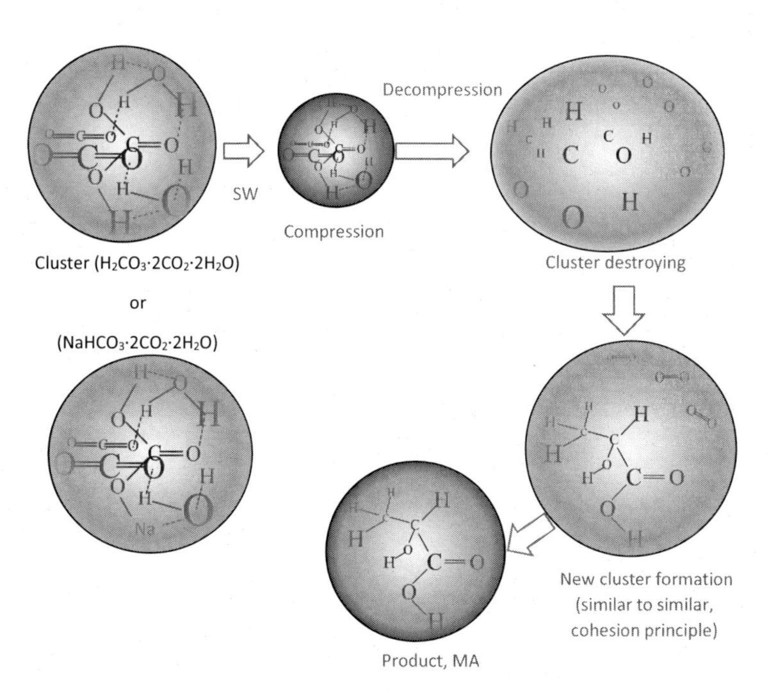

Figure 8. The idea of obtaining $H_6C_3O_3$ (MA) by the destruction of $H_2CO_3 \cdot 2CO_2 \cdot 2H_2O$ clusters in water under the influence of SW. The scheme of the reaction mechanism and it explanations was given in Scheme.

SW pulses weakly affect the main characteristics of the ensemble of domains and do not affect the base domain 1 RASC. However, they strongly affect the hexamer NC1, compacting it. Restoring the conformation of NC1 begins only 2 h after exposure to SW. This type of SW has a strong effect on 3RASC. The domain was greatly loosened and non-monotonous restoration of its shape also began only 2 hours after exposure to radiation.

According to the idea given in Table 5 and Figure 8, and Scheme (General idea, see below), the derivatives of H_2CO_3 and its soluble salts form clusters with the inclusion of water and CO_2. In this way, a concentrate of atomic nuclei is created which can be "made" under different conditions to form other combinations of nuclei - chemical compounds. Mild shock waves can become an instrument of compulsion in this case. The SW compacts the cluster, and the process of compressing it will temporarily destroy the chemical bonds. Further, the process of subsequent decompression will open up opportunities for new combinations of atomic links. The search for exposure conditions required finding not only capable hydrocarbons for this, but also conditions for the removal of oxygen, the preservation of organic compounds from secondary decay, and the supply of new portions of CO_2. Ideally, such a process would lead to the formation of hydroxypropionic acid and oxygen. The goal was to find the conditions for the formation of the $H_2CO_3{\cdot}2CO_2{\cdot}2H_2O$ cluster, compact it with the impact of the SW and create conditions for the rapid removal of gas and understand the transformation mechanism. It should be noted here that in the solution there are also individual clusters of carbonic acid like $(H_2CO_3)_6$. They can also be directly converted to glucose. What is the main process here? It requires additional research.

General idea: $H_2CO_3 + 2H_2O + 2CO_2 \rightarrow H_6C_3O_3 + O_2\uparrow$

and

$(H_2CO_3)_6 \rightarrow H_{12}C_6O_6 + 6O_2$□

By preliminary experiments, we founded that carbon dioxide and $NaHCO_3$ in water formed clusters (Figure 9) whose density was strongly affected by shock waves (Figure 10).

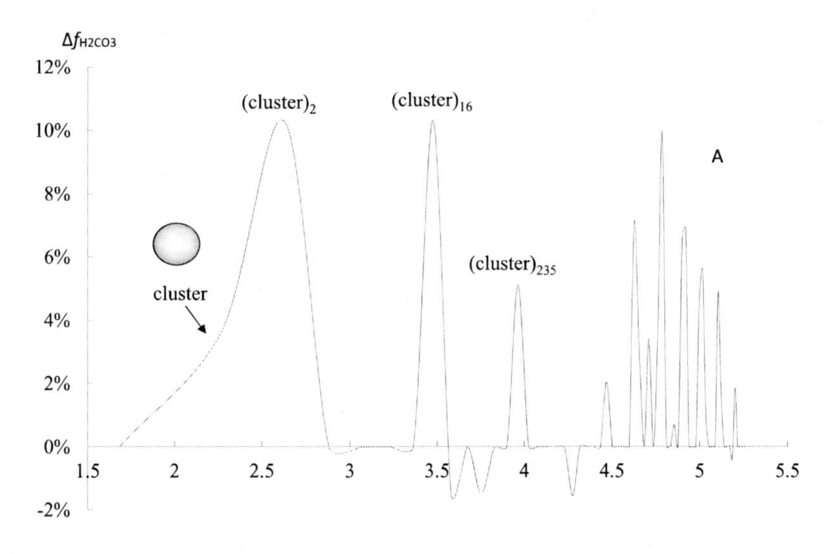

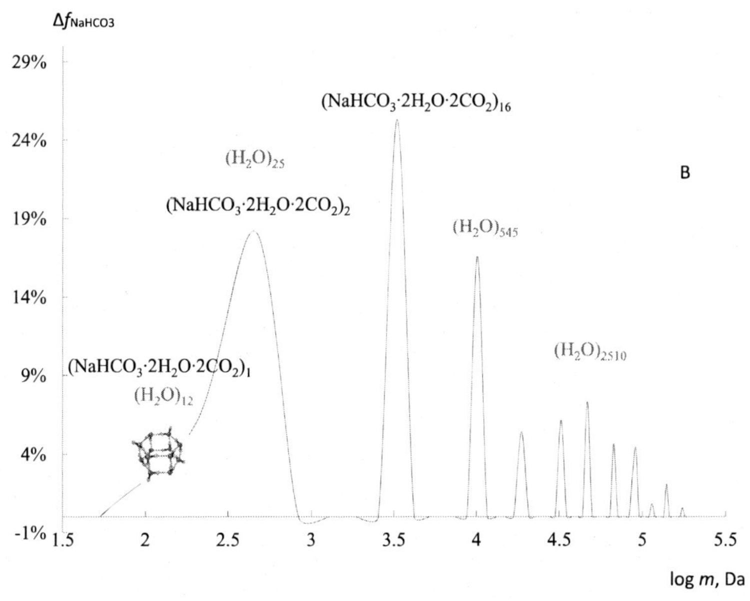

Figure 9. GMS spectra of a saturated solution of H_2CO_3 (A, 200 mg/l) in water and saturated $NaHCO_3$ in A (B) before being treated with shock waves. Here Δf was the difference between the values of f GN of the solution and the GN in the measuring cell. The structure of the base water cluster $(H_2O)_{12}$ [3] was kindly provided by Lenz [4].

It was noticed that when SW was exposed to an acid solution (A), gas was released into the solution. At the beginning of the irradiation process, the release of gas was the most intense, and by the end (15 min) it was almost finished. At the same time, the solution acquired a faint brown color. Similar was observed for the salt solution (B), however, in this case, the gas formation process was more uniform, and the separation of the organic product from the solution was possible only by additional salting out with saturated NaCl solution and neutralizing bicarbonates with HCl solution. At the same time, white flakes of a waxy substance appeared on the walls of the vessel and on the surface of the liquid. The product yield in terms of the H_2CO_3 solution ranged from 10 to 60 wt. %. LC (eluent water, SiO_2, 1 ml/min) showed the presence in A of at least 3 signals, difficultly identifiable substances (MA, lactides, possible Glucose and Glycerol too). The strongest and widest signal agreed with the MA signal. In case B, a wide spectrum of signals was detected. From the data of Figure 4A, it can be seen that, when SW was affected, the most active influence was on $(cluster)_2$, while on exposure to salt solution only the cluster $(NaHCO_3{\cdot}2H_2O{\cdot}2CO_2)_1$. So there were two close mechanisms for the synthesis of organic lactides (https://de.wikipedia.org/wiki/Polylactide) and MA (https://en.wikipedia.org/wiki/Lactic_acid).

In case A (Figure 9), we were unable to isolate the MA in a free form, the sample was very hydrophilic and all attempts to dry it were unsuccessful. Samples A and B burned without a trace, which clearly indicated their organic nature.

Let us briefly consider the mechanism of the impact of hydrocarbons on an ensemble of clusters in water, clusters in aqueous solutions of H_2CO_3 and its salts. The tables present the main characteristics of cluster ensembles in 3 liquids studied in this work. As the main characteristics of the ensembles, the number of cluster types (N) was taken before the SW exposure, during the SW energy pumping of the entire ensemble and the cluster ensemble reaction after the SW exposure stopped.

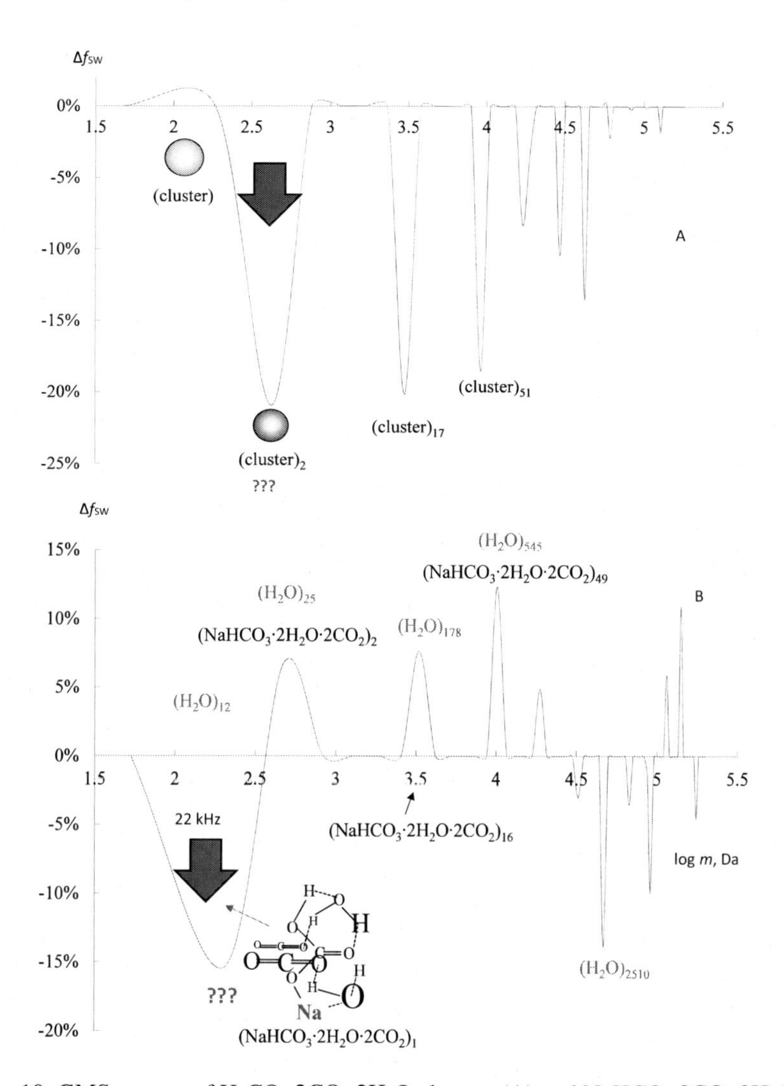

Figure 10. GMS spectra of $H_2CO_3{\cdot}2CO_2{\cdot}2H_2O$ clusters (A) and $NaHCO_3{\cdot}2CO_2{\cdot}2H_2O$ clusters in water (B). Zubow power constant 6E-15 N/m for A and 6.6E-15 N/m for B, GMS sensor - weak shock waves (<1 N/m^2), working shock waves from sound generator (20 kHz). A - the saturated solution CO_2 in water. B -the saturated solution of $NaHCO_3$ in a saturated solution of H_2CO_3. Both solutions were irradiation 15 minutes at 20 kHz. For case B, the formulas for water clusters were given over the signals, which were superimposed on the signals of the clusters $(NaHCO_3{\cdot}2H_2O{\cdot}2CO_2)_n$. ??? - the contribution to the intensity of the signal of the third compound formed by SW. Here, Δf_{SW} was the difference in the f_{SW} values for the H_2CO_3 solution after and before the SW exposure (A), for B the same, but for the salt solution.

Table 6. The main characteristics of cluster assemblies (up to 187 kDa) in distilled water (I), saturated aqueous H_2CO_3 solution (190 mg/l, II) and saturated $NaHCO_3$ solution in saturated aqueous H_2CO_3 solution (III). 20 kHz, 15 min. 390 K. Sine SW

I

H_2O	N	D_c	M_{GMS}, Da
fresh	11	0%	21251
1 min, at 20 kHz	27	10%	20968
15 min, at 20 kHz	26	7%	21412
1 min after irradiation	12	83%	25121
2 min after irradiation	13	22%	19168

II

$H_2CO_3 \cdot 2CO_2 \cdot 2H_2O$	N	D_c	M_{GMS}, Da
Fresh solution	13	19%	17521
1 min, at 20 kHz	24	1%	20728
15 min, at 20 kHz	24	3%	19954
1 min after irradiation	12	100%	17397
2 min after irradiation	12	96%	17433

III

$NaHCO_3 \cdot 2CO_2 \cdot 2H_2O$	N	D_c	M_{GMS}, Da
Fresh solution	13	0%	23209
1 min, at 20 kHz	28	16%	21661
15 min, at 20 kHz	29	15%	21730
1 min after irradiation	13	52%	52579
2 min after irradiation	13	61%	31659

As can be seen from the tables, the behavior of cluster ensembles in all the liquids studied as a whole was identical. However, there were differences. So the proportion of collapsed (dense) clusters (rich with energy) in the initial H_2CO_3 solution was significantly higher than in water and salt solution. The impact of hydrocarbons increased by more than 2 times the number of types of clusters and increased the contribution of dense structures only in water and salt solution. In a solution of an acid, on the contrary, their share has sharply decreased. On the other hand, the average molecular weight of the clusters responded poorly to the impact of hydrocarbons. Here, the phenomenon of a sharp increase in the proportion of dense clusters in all liquids after the end of SW exposure was interesting. It was especially pronounced for the acid (II) solution. It can be assumed that the forced pumping of hydrocarbons by the energy of an ensemble of clusters transferred it to a new thermodynamic state, and the discharge of energy led to a redistribution of energy content between the clusters in favor of the dominance of dense structures in the entire ensemble. This state was not stable and changes over time. Apparently this phenomenon also contributed to the chemical processes within the clusters and it was the strongest. It was possible to understand the mechanism of its impact from the standpoint of the wave process of mass restructuring in an ensemble, in which an internal SW is already formed inside the ensemble itself and its effect on chemical bonds was much stronger than the initial effect of the SW of a sound generator. The spectra in Figure 9 confirm this fact.

Conclusion

Weak shock waves of various shapes affect the long-range order in molecular matter in different ways.

There is a fundamental possibility to gently, selectively and purposefully influence the conformations of clusters in liquids and domains in polymers.

Inside the ensemble of mass clusters, the formation of strong shock waves during the distribution of energy between the clusters was possible.

The use of shock waves in chemistry and chemical technology opens a new period in the revolutionary development of natural science.

REFERENCES

[1] Zubow, K., Zubow, A. V., Zubow, V. A. *The Way to the ETIs. Applied gravitational mass spectroscopy*. Nova Sci. Publ. NY, 2014. https://www.novapublishers.com/catalog/product_info.php?products_id=42668&osCsid=5bd85d42dc273360fd48126de7be9daf.

[2] Zubow, K., Zubow, A., Zubow, V. A. Phenomenal properties of the domain ensembles in proteins. *Biochemistry & Molecular Biology Journal*. USA. 2016, vol. 2, no., 1, pp. 1-10.

[3] http://biochem-molbio.imedpub.com/phenomenal-properties-of-the-domain-ensembles-in-proteins.pdf.

[4] Bogdanov, E. V., G. M. Manturova Equicuster water model. *Biomedical Electronics*. 2000, № 7, p.19-28 (in Russian).

[5] Annika Lenz, Lars Ojamäe. On the stability of dence versus cage-shaped water clusters: Quantum-chemical investigations of zero-point energies, free energies, basis-set effects and IR spectra of $(H_2O)_{12}$ and $(H_2O)_{20}$. *Chem. Phys. Letters*. 2006, v. 418, pp. 361-367.

[6] Chaplin, M., SBU London, http://www.lsbu.ac.uk/water/index.

[7] Zubow, K., Zubow, A., Zubow, V. A. Zubow's method of utilization of carbon dioxide. New chemistry of shock waves. This is our a gift to humanity. *Advances in Chemistry Research*. 2019, vol. 57, pp. in print NY. https://novapublishers.com/shop/advances-in-chemistry-research-volume-57/.

[8] Shweta Singh, Sunil K. Srivastava and Dheeraj K. Singh. Hydrogen bonding patterns in different acrylamide-water clusters: microsolvatation probed by micro Raman spectroscopy and DFT calculation. *RSC Advances*. 2014, vol. 4, pp.1761-1774. DOI: 10. 1039/c3ra42707b.

[9] Bende, A., Perretta, G., Sementa, P., Di Palma, T. M. Inception of Acetic Acid/Water Cluster Growth in Molecular Beams. *A European Journal Chem. Phys. Chem. of Chemical Physics and Phsical Chemistry*. 2015 Oct 5. Vol. 16, no. 14, pp. 3021-3029. doi: 10.1002/ cphc.201500479. Epub. 2015 Aug. 20.

In: Horizons in World Physics. Volume 302 ISBN: 978-1-53616-472-5
Editor: Albert Reimer

Chapter 6

A CLOSER LOOK AT GLUONS

Sadataka Furui*
(formerly) Teikyo University,
Graduate School of Science and Engineering
Utsunomiya, Tochigi, Japan

Abstract

Gluons are strong interaction gauge fields which interact between quarks, i.e., constituents of baryons and mesons. Interaction of matters is phenomenologically described by gauge theory of strong, electromagnetic, weak and gravitational interactions. In electro-weak theory, left handed leptons l_L and neutrino ν_L, right handed leptons l_R and left handed quarks u_L, d_L and right handed quarks u_R, d_R follow $SU(2) \times U(1)$ symmetry. Charge of leptons and quarks define hypercharge Y, and via Higgs mechanism $SU(2)_L \times U(1)_Y$ symmetry forms $U(1)_{em}$ symmetry.

Presence of $J^P = 3/2^+$ baryons, or $N^{*++} \sim uuu$ suggests a new degree of freedom "color" for quarks, which follows $SU(3)$ symmetry group. Hence the gluon fields are expressed as $\mathcal{A}^a_\mu$ where $a = 1, 2, \cdots, 8$ specify the color $SU(3)$ bases, and μ are 4-dimensional space-time coordinates. The quantum electrodynamics was extended to quantum chromo dynamics (QCD). Since $\partial^\mu \mathcal{A}^a_\mu$ is not a free field, the gauge theory requires ghosts that compensates unphysical degrees of freedom of gluons. Gluons, ghosts, leptons and quarks are related by Becchi-Rouet-Stora - Tyuitin (BRST) transformation of electro-weak and strong interaction of the $U(1) \times SU(2) \times SU(3)$ symmetric Faddeev-Popov Lagrangian.

*Corresponding Author's E-mail: furui@umb.teikyo-u.ac.jp

In order to describe Hadron dynamics properly, embedding of 4-dimensional space to 5-dimensional space was tried in lattice simulations, and in light front holographic QCD (LFHQCD) approach in which conformally symmetric light-front dynamics without ghost are embedded in AdS_5, and a parameter that fixes a mass scale was chosen from the Principle of Maximum Conformality. Coulomb or Landau gauge fixed Faddeev-Popov Yang-Mills field equation is known to have the Gribov ambiguity, and tunneling between vacua between different topological structures was proposed by van Baal and collaborators. The symmetry of three colors can be assigned three vectors of quaternion $\mathbf{H}$, whose multiplication on 2×2 matrices of Dirac spinors on $\mathbf{S}^3$ induces transformations. Instantons or sphalerons whose presence is expected from conformal equivalence of $\mathbf{S}^3 \times \mathbf{R}$ to $\mathbf{R}^4$ are reviewed. An extension of the dynamics embedded in complex projective space is proposed for understanding field theories.

PACS 05.45-a, 52.35.Mw, 96.50.Fm.

Keywords: gluons, ghosts, BRST cohomology, Gribov-Zwanziger lattice simulation, LFHQCD, sphalerons, extra dimensions

1. INTRODUCTION

Feynman discusses on the problem of self-mass of charged particles in the section 28-4[1] that charges must be held to the sphere by some kind of rubber bands-something that keeps the charges from flying off. It was first pointed out by Poincaré [2] that the rubber bands -or whatever it is that holds the electron together-must be included in the energy and momentum calculation. He presented that the difference of the electromagnetic mass $m^{(1)} = U_{elec}/c^2$ where U_{elec} is calculated from $\epsilon_0 E^2$, where E is the electric field produced by the static charge, and $m^{(2)}$ calculated from momentum derived from the poynting vector $p = \frac{2}{3}\frac{e^2}{ac^2}\frac{v}{\sqrt{1-v^2/c^2}}$ are related by $m^{(2)} = \frac{4}{3}m^{(1)}$, and it was explained by Dirac[3] as effects of self energy of the charged particle. Feynman pointed out ambiguity in the self energy.

In order to study interaction of hadrons, solving Dyson-Schwinger equations [4, 5, 6] and lattice simulations [7] are two main methods. Dyson-Schwinger equation was used in analytical calculation of spectra of Quantum electro dynamics (QED), and applied to other fields. The lattice approach to quantum field theory shows evidence that gauge theory is the fundamental tool, and that exchange of gauge gluons can confine quarks within subnuclear matter. In comparison to electrons or muons which have $\mathbf{Z}_2$ symmetry $(1, -1)$, presence of three quark system $N^{*++} \sim uuu$ suggested quarks have $\mathbf{Z}_3$ symmetry which is called color symmetry (red, green, blue), but experimentally the symmetry is not detected.

In the minimal case of electromagnetism $\mathbf{Z}_2$ and gauge group $U(2)$[8], fermions which take values in the space $M_2 \otimes \mathbf{C}^2$, where M_n is a $n \times n$ dimensional matrix, the *Maxwell* action is given by

$$S_B = \frac{1}{4} \int F_{\alpha\beta} F^{\alpha\beta} dx,$$

and the *Dirac* action is defined by a set of vectors $\tilde{e}_a$ on $\mathbf{S}^2$ that satisfy $\tilde{e}_a \tilde{x}^b - C^b_{ca} \tilde{x}^c$, where $C_{abc} = r^{-1} \epsilon_{abc}$.

Dual to the basis $\tilde{e}_a$ are components $\tilde{\theta}^a$ of the Maurer-Cartan form:

$$d\tilde{x}^a = C^a_{bc} \tilde{x}^b \tilde{\theta}^c, \quad \tilde{\theta}^a = c^a_{bc} \tilde{x}^b d\tilde{x}^c - iA\tilde{x}^a.$$

Dirac proposed that the 1-form A contains the potential of *Monopole*[9, 10] of unit magnetic charge and $F = dA$.

$$S_F = Tr \int \bar{\psi}(\gamma^k D_k)\psi dx,$$

where $\gamma^k = (1 \otimes \gamma^\alpha, \sigma^a \otimes \gamma^5)$,

$$\gamma^0 = \gamma_0 = \begin{pmatrix} I & 0 \\ 0 & -I \end{pmatrix}, \quad -\gamma^k = \gamma_k = \begin{pmatrix} 0 & -\sigma_k \\ \sigma_k & 0 \end{pmatrix},$$

$$\sigma_1 = \begin{pmatrix} 0 & 1 \\ 1 & 0 \end{pmatrix}, \quad \sigma_2 = \begin{pmatrix} 0 & -i \\ i & 0 \end{pmatrix}, \quad \sigma_3 = \begin{pmatrix} 1 & 0 \\ 0 & -1 \end{pmatrix},$$

and

$$\gamma^5 = \frac{1}{4!} \epsilon_{\mu\nu\alpha\beta} \gamma^\mu \gamma^\nu \gamma^\alpha \gamma^\beta.$$

In the case of Yang-Mills field with connection ω and local section σ, the gauge transformation can be expressed as

$$\omega = h^{-1} A h + h^{-1} h$$

$$A' = \sigma'^{*}\omega = g^{-1}Ag + g^{-1}dg.$$

In terms of a moving frame on V

$$F = \frac{1}{2}F_{\alpha\beta}\theta^{\alpha} \wedge \theta^{\beta}, \quad D\psi = D_{\alpha}\psi\theta^{\alpha}.$$

Schwinger [6, 11] showed that there could exist a conserved divergenceless electromagnetic stress tensor, which guarantees stability and covariance of the theory. He considered spherically symmetric charge distribution of total charge e at rest,

$$\phi = ef(r^2), \quad \mathbf{A} = 0,$$

where $f(r^2) \sim (r^2)^{-1/2}$ and in uniformly moving rest frame

$$A^{\mu}(x) = \frac{e}{c}v^{\mu}f(\xi^2), \quad \xi^{\mu} = x^{\mu} + \frac{1}{c}v^{\mu}(\frac{1}{c}vx)$$

where $v^2 = v^{\mu}v_{\mu} = -c^2$ and $v\xi = 0$.

Differential of vector fields are

$$\partial^{\nu}A^{\mu}(x) = 2\frac{e}{c}\xi^{\nu}v^{\mu}f'(\xi^2)$$

and

$$\begin{aligned} \partial_{\mu}A^{\mu}(x) &= 2\frac{e}{c}\xi_{\mu}v^{\mu}f''(\xi^2) = 0 \\ F^{\mu\nu}(x) &= \partial^{\mu}A^{\nu}(x) - \partial^{\nu}A^{\mu}(x) = 2\frac{e}{c}(\xi^{\mu}v^{\nu} - \xi^{\nu}v^{\mu})f''(\xi^2). \end{aligned}$$

The current vector $j^{\mu}(x)$ produced by $\partial_{\nu}F^{\mu\nu} = \frac{4\pi}{c}j^{\mu}$ is

$$j^{\mu}(x) = \frac{e}{2\pi}v^{\mu}[-2\xi^2 f'''(\xi^2) - 3f'(\xi^2)]$$

which is the covariant form of the rest frame charge density

$$\rho = \frac{e}{2\pi}[-2r^2 f''(r^2) - 3f'(r^2)]$$

whose total charge $\int_0^{\infty} drr^2 4\pi\rho = e$ as calculated in[11].

The rest mass based on the covariant form and also on action-principle approach to electromagnetic mass: $m^{(1)}c^2$ and the electromagnetic mass based based on the Poynting vector $\frac{1}{4\pi}(\mathbf{E} \times \mathbf{B})$ defined as $m^{(2)}c^2$ are related by

$$m^{(1)}c^2 = \frac{3}{4}m^{(2)}c^2.$$

Presence of Poincaré stresses to explain the electromagnetic mass of an electron m_e, the inertial mass m_p and the effective mass $M = m_e + m_p$, that allow parametrization

$$Mc^2 = m_e c^2 + m_p c^2 = m_e c^2(1 + (1+h)/3) = m_e c^2(4/3 + h/3).$$

where parameter $h = -1$ yields electrostatic mass $m^{(2)}$, and the parameter $h = 0$ yields electrodynamic mass $m^{(1)}$ was proposed by Jackson [12]. The arbitraliness of the gauge fixing condition is discussed by [13, 14, 15, 16].

The standard model with Higgs mechanism [17, 18, 19, 20, 21] predicts that three of the original four $SU(2) \times U(1)$ gauge bosons become massive and one, corresponding to the photons, remains massless. Although Schwinger did not suppose stress tensors of non electromagnetic origin, gluons could play a role [12], since three quarks can have electromagnetic charge and affect charged particles.

Experimentally, Dirac's magnetic monopoles [9] are not observed, but 't Hooft [22] and Polyakov [23, 24] showed in the Yang-Mills field theory a string-like topological singularity can appear.

In section 2, we explain the gluon degrees of freedom and its symmetries in the Yang-Mills theory. When one fixes the gauge symmetry of gluons in 4-dimensional Yang-Mills field theory, it is necessary to include ghosts to compensate unphysical degrees of reedom of gluons. Taking eigenstates of ghost number, one can formulate BRST symmetry.

The experimental detection of Higgs boson which appear by spontaneous symmetry breaking of the vacuum pushed investigation of the structure of the vacuum. In section 3, we explain the Gribov ambiguity that appears in the gluon propagator, and incorporation of instanton and sphalerons in QCD, using quaternion bases $\mathbf{H}$, and embedding $\mathbf{S}^3 \times \mathbf{R}$ topology in $\mathbf{R}^4$ minkowski space.

In section 4, we present a few formulations of hadron dynamics with additional dimension which corresponds to chiral symmetry or symmetry under

Dirac's γ_5 operator. The first formulation consists of light-front holographic QCD (LFHQCD) by Brodsky and collaborators[26, 27, 32, 35] which is based on de Alfaro, Fubini and Furlan's anti de Sitter space approach [36]. The second formulation is based on non-commutative geometry developed by Connes and colaborators[25, 44, 45]. Non-commutative geometry is a set of tools to analyse broad range of objects that cannot be treated by classical methods [38, 50]. Non-commutative algebra appears when one incorporates Hamilton's quaternion $\mathbf{H}$ in the orthogonal bases of the Hilbert space[51, 55, 60]. Elements of $\mathbf{H}$ consist of $\mathbf{I}, \mathbf{i}, \mathbf{j}, \mathbf{k}$ which satisfy

$$\mathbf{ij} = \mathbf{k}, \quad \mathbf{jk} = \mathbf{i}, \quad \mathbf{ki} = \mathbf{j}$$
$$\mathbf{ji} = -\mathbf{k}, \quad \mathbf{kj} = -\mathbf{i}, \quad \mathbf{ik} = -\mathbf{j}$$

In the $SL(2, \mathbf{C})$ representation, we can set

$$\mathbf{I} = \begin{pmatrix} 1 & 0 \\ 0 & 1 \end{pmatrix}, \quad \mathbf{i} = \begin{pmatrix} 0 & i \\ i & 0 \end{pmatrix}, \quad \mathbf{j} = \begin{pmatrix} 0 & -1 \\ 1 & 0 \end{pmatrix}, \quad \mathbf{k} = \begin{pmatrix} i & 0 \\ 0 & i \end{pmatrix}.$$

and

$$\mathbf{A} = \mathbf{I}a + \mathbf{i}b + \mathbf{j}c + \mathbf{k}d = \begin{pmatrix} a + id & -c + ib \\ c + ib & a - id \end{pmatrix}, \quad a, b, c, d \in \mathbf{R}$$

and

$$\mathbf{A}^* = \mathbf{I}a - \mathbf{i}b - \mathbf{j}c - \mathbf{k}d.$$

We define $a + id = \alpha$ and $c + ib = \beta$, $\alpha, \beta \in \mathbf{C}$, and express

$$\mathbf{A} = \begin{pmatrix} \alpha & -\bar{\beta} \\ \beta & \bar{\alpha} \end{pmatrix}, \quad \mathbf{A}^* = \begin{pmatrix} \bar{\alpha} & -\beta \\ \bar{\beta} & \alpha \end{pmatrix}.$$

From algebra of complex numbers and quaternions, one can construct Clifford algebra [55] in quadratic spaces. In section 5, we explain Clifford Algebra on the manifold of $\mathbf{S}^3$ and embed $\mathbf{S}^3 \times \mathbf{S}^1$ topology to complex projective space $\mathbf{CP}^3$.

Discussion on symmetry violation and topological approach in other physical phenomena is given in section6.

2. GLUONS AND BRST SYMMETRY OF YANG-MILLS FIELD

In Hamiltonian formulation of Yang-Mills field [9, 70], lagrangian of QCD in the first order is written as

$$\mathcal{L} = -\frac{1}{2}tr[\mathcal{E}_k\partial_0\mathcal{A}_k - \frac{1}{2}(\mathcal{E}_k^2 + \mathcal{G}_k^2) + \mathcal{A}_0\mathcal{C}], \tag{1}$$

where

$$\begin{aligned}\mathcal{E}_k &= \mathcal{F}_{k0}, \mathcal{G}_k = \frac{1}{2}\epsilon^{ijk}\mathcal{F}_{ji}, \mathcal{C} = \partial_k\mathcal{E}_k - g[\mathcal{A}_k, \mathcal{E}_k] \\ \mathcal{F}_{ik} &= \partial_k\mathcal{A}_i - \partial_i\mathcal{A}_k + g[\mathcal{A}_i, \mathcal{A}_k].\end{aligned}$$

The gauge fields $\mathcal{A}_k$ containing the bases λ^a of $SU(3)$ are generators of the symmetry, and they are decomposed in longitudinal and transverse comoponents

$$\mathcal{A}_k = \mathcal{A}_k^L + \mathcal{A}_k^T$$

where $\partial_k\mathcal{A}_k^T g = 0$ and

$$\partial\mathcal{A}_k^L = \partial_k\mathcal{B}(x), \quad \mathcal{B}(x) = \frac{1}{4\pi}\int\frac{1}{|x-y|}\partial_k\mathcal{A}_k(y)dy.$$

When the transverse component $\mathcal{E}_k^T(x)$ is defined as $\partial_k Q(x)$, the constraint on the longitudinal component is written as

$$\Delta Q - g[\mathcal{A}_k, \partial_k Q] - g[\mathcal{A}_k, \mathcal{E}_k^T] = 0.$$

In the gauge $\mathcal{A}_0 = 0$, three dimensional expression of $\mathcal{F}_{ij}$ becomes

$$\mathcal{F}_{ik} = \partial_k\mathcal{A}_i - \partial_i\mathcal{A}_k + g[\mathcal{A}_i, \mathcal{A}_k].$$

Madore[8] defined orthogonal moving frame θ^α, whose metric satisfy

$$g(\theta^\alpha \otimes \theta^\beta) = g^{\alpha\beta}$$

and structure equations of the torsion form Θ^α and curvature form Ω^α_β defined as

$$\begin{aligned}\Theta^\alpha &= d\theta^\alpha + \omega^\alpha_\beta \wedge \theta^\beta \\ \Omega^\alpha_\beta &= d\omega^\alpha_\beta + \omega^\alpha_\gamma \wedge \omega^\gamma_\beta\end{aligned}$$

satisfy Bianchi identity:

$$d\theta^\alpha + \omega^\alpha_\beta \wedge \Theta^\beta = \Omega^\alpha_\beta \wedge \theta^\beta$$
$$d\Omega^\alpha_\beta + \omega^\alpha_\gamma \wedge \Omega^\gamma_\beta - \Omega^\alpha_\gamma \wedge \omega^\gamma_\beta = 0.$$

Gauge transformations of Yang-Mills tensor fields are performed by using the covariant derivative D in $\mathbf{R}^n$ [8, 57, 59]

$$S[A] = \frac{1}{2}Tr\int F \wedge *F = \frac{1}{4}Tr\int F_{\alpha\beta}F^{\alpha\beta}dx$$

where Hodge duality map $*$ of $\Omega^p(V)$ into $\Omega^{n-p}(V)$ is defined for $p \leq n$ to the $(n-p)$ forms as

$$*\theta_{\alpha_1\cdots\alpha_p} = \frac{1}{(n-p)!}\epsilon_{\alpha_1\cdots\alpha_p\alpha_{p+1}\cdots\alpha_n}\theta^{\alpha_{p+1}} \wedge \cdots \wedge \theta^{\alpha_n}.$$

The vacuum Yang-Mills equations $D_\alpha F^{\alpha\beta} = 0$ where

$$F = \frac{1}{2}F_{\alpha\beta}\theta^\alpha\theta^\beta$$

satisfies the gauge transformation relations $F' = g^{-1}Fg$. If the gauge group is abelian, the equatio is the Maxwell equation. When the gauge group is non-abelian, the field eqautions become non-linear, and non-perturbative effects become important.

Quarks and gluons have $SU(3)$ color degrees of freedom, but these degrees of freedom are not observed experimentally and models of color confinement were proposed. Functionals of the QCD green function are given by Lagrangian $\mathcal{L}$ describing quarks and gluons, gauge-fixing Lagrangian $\mathcal{L}_{Fix}$ and Faddeev-Popov ghost Lagrangian $\mathcal{L}_{Ghost}$.

Faddeev-Popov ghosts are unphysical fields that compensate unphysical gluon degrees of freedom [21, 68, 69, 71, 78].

$$\begin{aligned}
\mathcal{L} &= \frac{1}{4}F^a{}_{\mu\nu}(x)F^{\mu\nu}{}_a(x) + \bar\psi(x)(i\gamma^\mu D_\mu - M)\psi(x), \\
\mathcal{L}_{Fix} &= -\frac{1}{2}(C^a[\mathcal{A}_\mu])^2 = -\frac{1}{2\xi}(\partial^\mu A_\mu{}^a(x))^2, \\
\mathcal{L}_{Ghost} &= \bar u^a(x)\partial^\mu(\partial_\mu\delta_{ab} - gf_{abc}A_\mu{}^c(x))u^b(x).
\end{aligned} \tag{2}$$

The Becchi-Rouet-Stora-Tyutin (BRST) transformation in infinitesimal form is[84, 68]

$$\begin{aligned}
\delta_s A^a_\mu(x) &= -f^{abc} A_\mu{}^b(x) u^c(x) \delta\bar\lambda - \frac{1}{g}\partial_\mu u^a(x)\delta\bar\lambda \\
\delta_s \psi(x) &= -iT_a \psi(x) u^a(x)\delta\bar\lambda \\
\delta_s \bar\psi(x) &= i\bar\psi(x) T_a u^a(x)\delta\bar\lambda \\
\delta_s u^a(x) &= -\frac{1}{2} f^{abc} u_b(x) u_c(x)\delta\bar\lambda \\
\delta_s \bar u^a(x) &= \frac{1}{g\sqrt{\xi}} C^a[A_\mu(x)]\delta\bar\lambda = \frac{1}{g\xi}\partial^\mu A_\mu{}^a(x)\delta\bar\lambda
\end{aligned} \tag{3}$$

where ξ is the gauge parameter.

The covariant derivative of ψ is with color indices k, i and $SU(3)$ color generator t^C,

$$(D_\alpha\psi)_k = [\partial_\alpha \delta_k{}^l - igA_\alpha{}^C (t^C)_k{}^l](\psi)_k.$$

Henneaux and Teitelboim [71] defined the phase space P and smooth phase space function $C^\infty(P)$. One restricts P a surface Σ and functions that vanish on Σ form an ideal in $C^\infty(P)$ which is denoted as $\mathcal{N}$.

The local Lagrangian L and action on variables y^i are defined by

$$S[y(t)] = \int_{t_1}^{t_2} L(y^i, \dot y^i, \cdots y^{(k)_i}) dt$$

where $y^{(k)_i} = d^k y_i/dt^k$, Gauge transformations of arbitrary parameter ϵ are

$$\delta_\eta y^i = S_A{}^i \eta^A$$

$$\delta_\epsilon S = \frac{\delta S}{\delta y^i}\delta_\epsilon y^i = \frac{\delta S}{\delta y^i} R^i_\alpha \epsilon^\alpha$$

and Noether's identity is

$$\frac{\delta S}{\delta y^i} R^i_\alpha = 0.$$

Kugo and Ojima [77, 78] pointed out that the QCD Lagrangian is invariant under the BRS transformation, and the gauge fields $A_\mu{}^a(x)$, the auxiliary field $B^a(x)$, the covariant derivative of the ghost field $D_\mu c(x)$, anti-ghost field $\bar c(x)$

necessarily have massless asymptotic fields which form the BRS-quartet [79]. The physical space $\mathcal{V}_{phys} = |phys\rangle$ is specified as the one that satisfies the condition $Q_B|phys\rangle = 0$, where

$$Q_B = \int d^3x[B^a D_0 c^a - \partial_0 B^a \cdot c^a + \frac{i}{2} g\partial_0 \bar{c}^a \cdot (c \times c)^a]$$

and $(F \times G)^a = f_{abc} F^b G^c$.

The Noether current[71] corresponding to the conservation of the color symmetry is

$$gJ_\mu{}^a = \partial^\nu F_{\mu\nu}{}^a + \{Q_B, D_\mu \bar{c}\},$$

where the ambiguity by divergence of antisymmetric tensor shuld be understood, and this ambiguity is utilised so that massless contribution may be eliminated for the charge Q_B to be well defined [111, 112]. Denoting $g(A_\mu \times \bar{c}) \to u^a{}_b \partial_\mu \bar{\gamma}^b$, one obtains A has a vanishing asymptotic value and

$$\int e^{ip(x-y)} \langle 0|TD_\mu c^a(x) g(A_\nu \times \bar{c})_b(y)|0\rangle dx = (g_{\mu\nu} - \frac{p_\mu p_\nu}{p^2}) u^a_b(p^2)$$

where T means Dyson's time-ordering product operations that appear in perturbative Schroedinger functional approach[4].

Kugo and Ojima[78] modified the Noether current for color charge Q^a such that

$$gJ'^{\,a}_\mu = gJ_\mu - \partial^\nu F_{\mu\nu}{}^a = \{Q_B, D_\mu \bar{c}\}^a$$

and the condition for the massless component in the current $\{Q_B, D_\mu \bar{c}\}$ is absent is $\delta^b_a + u^b_a = 0$.

In Landau gauge, Gribov region Ω is specified by the variation with respect to $g = e^\epsilon$

$$\begin{aligned} \Delta||A^g||^2 &= -2\langle \partial A|\epsilon\rangle + \langle \epsilon \partial \mathcal{D}|\epsilon\rangle \\ \Omega &= \{A| - \partial \mathcal{D} \geq 0, \partial A = 0\}. \end{aligned}$$

We defined lattice link operator $U_{x\mu} = e^{A_{x,\mu}}$, where $A_{x,\mu} = -A^\dagger_{x,\mu}$ is the $SU(3)$ Lie algebra operator, and gauge transformation was chosen as

$$e^{A^g_{x,\mu}} = g^\dagger_x e^{A_{x,\mu}} g_{x+\mu}.$$

Landau gauge was realized by choosing $g = e^{\epsilon}$ and minimizing

$$||A^g||^2 = \sum_{x,\mu} Tr A^g_{x,\mu}{}^{\dagger} A^g_{x,\mu}.$$

The lattice covariant derivative $D_\mu(A) = \partial_\mu + Ad(A_\mu)$ in SU(3) simulation is given in [111].

Zwanziger[85, 86] developed a globally correct gauge fixing procedure. He showed in the of minimal Landau or Coulomb gauge lattice simulation, gluon propagator is infrared vanishing

The Faddeev-Popov determinant for gauge fixed fields A, which are hermitian Lie algebra gauge fields [53, 54, 121, 133] is

$$FP(A) = -\partial_i D_i(A) = -\partial_i(\partial_i + iadA_i)$$

where $(adX)Y = [X, Y]$.

Using notations of [92] and compactified vector fields $M \sim \mathbf{S}^3$, Coulomb gauge condition $\partial_i A^i = 0$ is realized at critical points of

$$I(g; A) = \int_M Tr(\{[g]A_i\}^2) = \int_M Tr(\{A_i + ig^{-1}\partial_i g\}^2).$$

For $I(hg; A) = I(h; [g]A)$ with $h = e^X$ and using

$$e^{-X}\partial_i e^X = \frac{1 - exp(-adX)}{adX}(\partial_i X) = \partial_i X + \frac{1}{2}[\partial_i X, X] + \frac{1}{6}[[\partial_i X, X], X] + \cdots,$$

one finds

$$I(e^X; A) = ||A||^2 - 2i\int_M Tr(X\partial_i A_i) + \int_M Tr(X^\dagger FP(A)X)$$

$$+\frac{i}{3}\int_M Tr(X[[A_i, X], \partial_i X]) + \cdots.$$

The Gribov region Ω is defined as the set of transverse gauge fields for which $FP(A)$ is positive. By gauge transformation, a trajectory of minimum points of $FP(A)$ could bifurcate into two trajectories of minimum points [95, 96]

Zwanziger [97] defined fundamental modular region Λ specified by

$$\Lambda = \{A|||A||^2 = Min_g||A^g||^2\}, \quad \Lambda \subset \Omega$$

and the ghost propagator which is an inverse of the Faddeev-Popov operator $\partial\mathcal{D}$[72] as

$$G(x-y)\delta^{ab} = \langle \frac{1}{-\partial\mathcal{D}_\mu(A)} \rangle.$$

He defined lattice gauge covariant derivative D_μ^{ab} that operates on link variable $U_\mu(x)$ whose relation to $A_\mu^a(x)$ is

$$A_\mu^a(x)t^a = \frac{1}{2}[U_\mu(x) - U_\mu^\dagger(x)]_{traceless}.$$

In the 't Hooft double line representation [75], quarks are described by single line, gluons and ghosts are described by double line. The Faddeev-Popov tensor gives the ghost propagator

$$G_{\mu\nu}(x-y)\delta^{ab} = \langle D_\mu^{ac} D_\nu^{bd} \frac{1}{-\partial\mathcal{D}(A)^{cd}} \rangle$$

and its Fourier transform is

$$G_{\mu\nu}(\theta) = \sum_x G(x)_{\mu\nu} e^{-i\theta(x+\frac{1}{2}e_\mu - \frac{1}{2}e_\nu)}$$

whose projection in the transverse direction is

$$G_{\mu\nu}^{TT} = (g_{\mu\nu} - \frac{p_\mu p_\nu}{p^2})G_{\mu\nu}.$$

The ghost number correspods to the number of branching of a gluon in double line expressions. Infrared vanishing of the gluon propagator predicted by Zwanziger [86] is consistent with the renormalizable theory of 't Hooft [126].

Baulieu and Zwanziger [104] replaced the gauge fixing in 4 dimensional space to 5-dimensional formulation and succeeded in restricting 4 dimensional gluon field in the Gribov region and decoupling ghost fields. Zwanziger[105] derived an equation similar to Kugo-Ojima confinement equation.

Schaden and Zwanziger (SZ)[106, 107] introduced change of variables of BRST ghost pairs $\phi^a_{\nu b}$, $\bar{\phi}^a_{\nu b}$, $b^a = s\bar{c}$, and $\bar{c}^a$ as

$$\begin{aligned} \phi^a_{\nu b}(x) &= \varphi^a_{\nu b}(x) - \gamma^{1/2} x_\nu \delta^a_b \\ \bar{\phi}^a_{\nu b}(x) &= \bar{\varphi}^a_{\nu b}(x) + \gamma^{1/2} x_\nu \delta^a_b \\ \hat{b}^a(x) &= b^a(x) + i\gamma^{1/2} x_\nu f^a[\bar{\varphi}_\nu(x)] \\ \hat{\bar{c}}^a(x) &= \bar{c}^a(x) + i\gamma^{1/2} x_\nu f^a[\bar{\omega}_\nu(x)], \end{aligned} \quad (4)$$

where a constant γ was defined by a gauge fixing condition, and produced generators $M_{\mu\nu}$ and P_μ, that satisfy the Poincaré group algebra, using A_μ φ_μ, $\bar{\varphi}_\mu$ and ghost fields. They showed that BRST charge Q_B does not commute with P_μ and the theory cnnot be invariant under both translation and BRST. Usual Poincaré group algebra consists of Lorentz generator, and Coleman-Mandula theorem [108] says that the algebra of internal symmetry or the BRST symmetry cannot carry space-time indices like x. SZ's Poincaré generators are different from usual ones, but they found that the Yang-Mills equation obtained by the Gribov-Zwanziger prescriptions are not unique, due to the gauge freedom of A. Coleman-Mandula theorem does not apply for supersymmetric system. The Haag-Lopuszanski-Sohnius theorem [109] says that the conditions on internal symmetry like BRST symmetry are similar in supersymmetric system [38]. SZ claim that the ground state energy obtained by choosing the fundamantal modular region is unique, and preservation of the BRST symmetry is equivalent to that of the Kugo-Ojima confinement criterion.

Although the BRST symmetry is violated on lattice simulations [103], samples that we took close to the fundamental moduler region showed $u \sim -0.7$ in quenched simulation [112, 114], and $u \sim -1$ in unquenched Kogut-Susskind fermion [82, 83] configurations produced by the MILC collaboration [115]. Whether the BRST symmetry is preserved in large lattices remains open.

Verification of Kugo-Ojima coefficient using Dyson-Schwinger equation was investigated by Alkofer et al. [99, 100, 101].

Modern formulation of the Faddeev-Popov S-matrix elements of Yang-Mills field theory in Landau gauge is

$$\langle in|out\rangle \sim \int exp^{iS[B]}\Delta[B]\Pi_x\delta(\partial_\mu B^\mu(x))dB(x)\int \Pi_x D\Omega(x)$$

where the gauge group is expressed as

$$B_\mu \to B_\mu^\Omega = \Omega B_\mu \Omega^{-1} + \epsilon^{-1}\partial_\mu \Omega \Omega^{-1}$$

and the factor $\Pi_x \delta(\partial_\mu B^\mu(x))$ symbolyzes that integral is performed over transverse fields, and $\Delta[B]$ is chosen such that

$$\Delta[B] \int \Pi_x \delta(\partial_\mu B^\mu(x)^\Omega) d\Omega = const$$

holds. The analysis was extended from the real Minkowski space to the complex Riemann space in [61], and a selection of physically relevant curves in the gauge transformed orbit space was done using fiber bundles [66].

3. The Gribov Copy Problem and the Structure of Vacuum

Serious ambiguities in Coulomb or Landau gauge fixed Faddeev-Popov Yang-Mills field equation were studied in connection with BRS transformation [73, 74]. Gribov ambiguity and gauge fixing procedures are investigated in [76]. Spontaneous symmetry breaking of the gauge theory and prediction of a scalar bosons [17, 18, 19, 20] motivated studies of topology of Higgs field [124].

Polyakov [24] proposed instanton in non abelian gauge theory, Dashen, Hasslacher and Neveu [87] and Yaffe [88] proposed sphalerons, which in $\mathbf{S}^3 \times \mathbf{R}$ described by

$$\begin{aligned} F_{ij} &= \partial_i A_j - \partial_j A_i + 2\epsilon_{ijk} A_k + [A_i, A_j] \\ &= \frac{2i\epsilon_{ijk}\tau_k s^2}{(1+s^2+b^2+2sb\cdot n)^2}, \end{aligned}$$

where the gauge vectors of instantons are

$$A_0 = \frac{is\vec{b}\cdot\vec{\tau}}{1+s^2+b^2+2sb\cdot n}, \quad A_i = -i\frac{s^2+sb\cdot n)\tau_j + s(\vec{b}\wedge\vec{\tau})_j}{1+b^2+s^2+2sb\cdot n},$$

where $\vec{b} = b\cdot\vec{e}$, e_μ^i is the dreibein and $s = \lambda e^t$.

The sphaleron potential is

$$\mathcal{V} = -\frac{1}{2}\int_{S^3} Tr(F_{ij}^2) = \frac{48\pi^2(1+s^2+b^2)s^4}{(1+s^2+b^2)^2 - 4s^2b^2)^{5/2}}.$$

Take any path γ connecting $A_i = 0$ and $A_i = n\cdot\sigma\partial_i n\cdot\bar{\sigma}$ and determining maximum $\mathcal{V}_m(\gamma)$, one defines the sphaleron [94, 102].

A spharleron have a definite Chern-Simons number

$$Q(A) = \frac{1}{8\pi^2}\int_{S^3} Tr(A\wedge dA + \frac{2}{3}A\wedge A\wedge A)$$

in $SU(2)$-Higgs theory [119]. Higgs boson of a mass of 125 GeV/c^2 was detected at Large Hadron Collider at CERN, between 2011 and 2012.

For a $g \in \mathcal{G}$, where $\mathcal{G}$ is gauge configurations, $F_A(g) = I(g;A)$ is for generic A, Morse function [89] on $\frac{g}{\mathcal{G}}$, where g is the group of local gauge transformation [92]. Properties of $U(n)$ anti-self-dual connections A on a C^n bundle over a torus T^4 and relation to instantons i.e., self-dual solutions of the Yang-Mills equation [91] were studied by Braam and van Baal [90]. van Baal and Hari Dass [94] studied instanton and sphaleron in the $SU(2)$ Yang Mills gauge theory. They considered $x_\mu \in \mathbf{R}^4$, $r^2 = x_\mu^2$ and $n_\mu = x_\mu/r$ and redefined time t through $r = Rexp(t/R)$ such that

$$dx_\mu^2 = exp(2t/R)(dt^2 + R^2 dn_\mu^2)$$

where dn_μ^2 represents the metric of a unit sphere of volume $2\pi^2$, and coordinates x_μ are represented by quaternions [52, 53, 55, 56] $\sigma_m u = (\mathbf{I}, i\tau)$ and anti-quaternion $\bar{\sigma}_m u = (\mathbf{I}, -i\tau)$ as

$$x = x_\mu\sigma^\mu = \bar{x}^\dagger, \quad \sigma_i = -\bar{\sigma}_i = i\tau_i, \quad \sigma_4 = \bar{\sigma}_4 = 1.$$

The t dependence of solutions follows from stability of the vacuum. Effects of finite temperature is studied in [120].

Choosing $R = 1$ and defining dreibein on $\mathbf{S}^3$, using ‘t Hooft symbols η and $\bar{\eta}$ [125]

$$\begin{aligned}\sigma_\mu\bar{\sigma}_\nu - \sigma_\nu\bar{\sigma}_\mu &= 2i\eta_{\mu\nu}^a\tau_a \\ \bar{\sigma}_\mu\sigma_\nu - \bar{\sigma}_\nu\sigma_\mu &= 2i\bar{\eta}_{\mu\nu}^a\tau_a\end{aligned}$$

where $a = 1, 2, 3$ and μ, ν run from 1 to 4, as

$$e^a_\mu = \eta^a_{\mu\nu} n_\nu.$$

In the absence of fermions, the θ parameter is the relevant quantum number to connect the wave functionals in te various vacua. van Baal and Cutkosky [93] verified the tunnelling from the $\mathbf{A} = 0$ vacuum to the vacua that have Chern-Simons number one or minus one in $SU(2)$ gauge.

$\mathbf{S}^3 \times \mathbf{R}$ is isomorphic to torus $T^4 = \mathbf{C}^4/G$ where G is a discrete subgroup of a complex manifold W.

4. Hadron Dynamics without Ghosts in Space-Time with an Extra Dimension

'‘t Hooft and Veltman [122] pointed out that in 4-dimensional space, one loop spinor diagram with the vertex

$$\gamma^5 = \frac{1}{4!}\epsilon_{\mu\nu\alpha\beta}\gamma^\mu\gamma^\nu\gamma^\alpha\gamma^\beta$$

contributes anomalies due to singularities of Feynman integral, but they can be cancelled to make the theory renormalizable [123]. The chiral flavor $SU(N)_L \otimes SU(N)_R \otimes U(1)$ currents

$$\begin{aligned} J^{st}_\mu &= i\bar{\psi}^s\gamma_\mu\psi^t \\ J^{5st}_\mu &= i\bar{\psi}^s\gamma_\mu\gamma_5\psi^t \end{aligned}$$

are conserved but

$$J^5_\mu = \sum_t J^{5tt}_\mu$$

has the Adler-Bell-Jackiw anomaly charcterized by

$$\begin{aligned} \partial_\mu J^5_\mu &= -i(Ng^2/16\pi^2)G^a_{\mu\nu}\tilde{G}^a_{\mu\nu} \\ G^a_{\mu\nu} &= \partial_\mu A^a_\nu - \partial_\nu A^a_\mu + g\epsilon_{abc}A^b_\mu A^c_\nu \\ \tilde{G}^a_{\mu\nu} &= \frac{1}{2}\epsilon_{\mu\nu\alpha\beta}G^a_{\alpha\beta} \end{aligned}$$

and topological quantum number

$$n = (g^2/32\pi^2) \int G^a_{\mu\nu} \tilde{G}^a_{\mu\nu} d^4x.$$

't Hooft linked isospin to one of $SO(3)$ subgroups of $SO(4)$ [125].

Dosch, de Téramond and Brodsky [28] showed that if one embeds 4-dimensional Euclidean space in 5-dimensional anti-de Sitter space [129, 130] by adopting the AdS/CFT corespondence [131], and choosing the light-front holographic coordinate, one can construct QCD theory without ghost and get insight to the color confinement problem.

Conformal supergauge transformations in 4-dimension were studied by Wess and Zumino [49]. They pointed out that in two dimensions, supergauge transformation is connected to the absence of ghost states. Their gauge transformations in 4 dimension turned out to be a combination of a conformal transformation and a γ_5 transformation.

The supersymmetry of the low energy hadron spectra was proposed in [36, 37], and applied in the Principle of Maximum Conformality (PMC), which was successful in explaining supersymmetric hadron spectroscopy [29, 34] and infrared suppression of gluon propagators. In light-front quantum field theory, Pauli-Lubanski pseudovector [38],

$$W^\mu = -\frac{1}{2}\epsilon^{\mu\nu\alpha\beta} P_\nu M_{\alpha\beta}$$

where $P^2 = m^2$ and W^2 are Casimir operators, and one can use light-front gauge in all Lorentz frames avoiding redundant gauge degrees of freedom characteristic of covariant gauges [32]. Maximum Conformal Light Front Holographic QCD has relations to twistor methods [129, 130].

In the holographic theory of Susslind and Lindesay [132], there are two two observers "Fido" and "Frefo". The latter is a freely falling observer in gravitational fields. When the worlds observed by Frefo is Lorentz transformed, the world observed by Fido should appear. In estimation of electromagnetic radiation of electrons with the velocity close to that of light (Frefo), the Weizsäcker-Williams approximation [141] was adopted from the data of static electros(Fido). Similar principle can be applied to QCD.

Based on chiral symmetry of $U_A(1)$ and $SU(N_f)$-flavor symmetry, interaction of quars with gluons in backgrounds of instantons at finite temperature was studied in [39], and a non-linear mesonic σ model in which baryons are treated as Skyrme chiral solitons was proposed by Chemtob [40, 41]. Following Witten [42, 43], 4-dimensional space-time M was embedded in 5-dimensional space-time Q as $M = \partial Q$, described by coordinates $y^i = (t, \vec{x}, \rho)$, and Wess-Zumino-Witten action functional $A(t)$ on the time circle $S^1 = [0, 2\pi]$ was extended on a disk (t, ρ) bounded by S^1, as $A(t, \rho) \in SU(3)/U(1)$.

Separation of quark parts and gluon parts of nucleon angular momentum tensor represented by the Pauli-Lubanski pseudo-vector, and similarity of the formalism to light front dynamics was discussed by Ji [127, 128]. He defined energy momentum of a nucleon $T^{\mu\nu} = T_q^{\mu\nu} + T_g^{\mu\nu}$ and total momentum operator $P^\mu = \int d^3x T^{0\mu}$ The helicity operator $h = \vec{J} \cdot \hat{P}$ where $\hat{p} = \vec{p}/|\vec{p}|$ coincides with $W^\mu s_\mu$, where $s_\mu = (|\vec{p}|/M, p^0\hat{p}/M)$.

The choice of light-front time $\tau = t + z/c$ and light-cone coordinates $\xi_\pm = (\xi_0 \pm \xi_3)/\sqrt{2}$, and the transverse component $\xi_\perp = (\xi_1, \xi_2)$ allows Lorentz transformation proper conformal analysis.

$$\langle p|P^\mu_{q,g}|p\rangle = A_{q,g}(\mu)p^\mu p + B_{q,g}(\mu)g^{\mu 0}/(2p^0).$$

The angular momentum density $M^{\mu\alpha\beta} = T^{\mu\beta}x^\alpha - T^{\mu\alpha}x^\beta$ consists of quark and gluon parts

$$M^{\mu\alpha\beta}_{q,g} = T^{\mu\beta}_{q,g}x^\alpha - T^{\mu\alpha}_{q,g}x^\beta.$$

Gluon contribution to the angular momentum is

$$J^i_g = \frac{1}{2}\epsilon^{ijk}\int c^3x(T^{0k}_g x^j - T^{0j}_g x^k)$$

where

$$T^{\mu\nu}_g = \frac{1}{4}g^{\mu\nu}F^2 - F^{\mu\alpha}F^\nu_\alpha,$$

and $\vec{J}_g$ is expressed by the Poynting vector as $\vec{J}_g = \int d^3x(\vec{E} \times \vec{B})$.

In non-commutative geometry, one considers $n-$dimensional Euclidean vector space $\mathbf{C}^n$ with inner product, and interactions are descibed by $n \times n$ complex matrices M_n. Let V be a smooth compact oriented real manifold without boundary of dimension m, and $C(V)$ be the commutative associative algebra of smooth real-valued function on V. Let ∂_i be the natural basis of vectors on the space $\mathbf{R}^n$ and linear combination of $X = X^i\partial_i$ with $X^i \in \mathcal{C}(\mathbf{R}^n)$ modules and

$$\chi(\mathbf{R}^n) = \oplus_1^n \mathcal{C}(\mathbf{R}^n).$$

Maxwell-Dirac action [8] can be defined by the algebra $\mathcal{A} = \mathcal{C}(V) \otimes M_n$ where M_n is the matrix algebra.

One lets $\mathbf{C}^n$ be the n-dimensional Euclidean vector space with the standard inner product, and for an integer $m < n$ write $\mathbf{C}^n = \mathbf{C}^m \oplus \mathbf{C}^{n-m}$, and decompose M_n as a direct sum $M_n = {M_n}^+ \oplus M_n^-$. The M_n^+ are even operators that satisfy $M_n^+ = M_m \times M_{n-m} \subset M_n$.

Yang-Mills action for $n = 3$ and $m = 1$ can be defined in 3 dimensional even matrices

$$\mathbf{H} \oplus \mathbf{C} \subset {M_3}^+$$

The total algebra is chosen to be $\mathcal{A} = \mathcal{C}(V) \otimes {M_3}^+$ and modules of 1-form is

$$\Omega^1(\mathcal{A}) = \mathcal{C}(V) \otimes {\Omega^1}_\eta \oplus \Omega^1(\mathcal{C}(V)) \otimes {M_3}^+$$

where $\Omega^1(\mathcal{C}(V))$ is the module of de Rham forms [8]. ${\Omega^1}_\eta$ is ${M_3}^-$, i.e., polynomials of odd matrices of type

$$\eta = \begin{pmatrix} 0 & 0 & a_1 \\ 0 & 0 & a_2 \\ -a_1^* & -a_2^* & 0 \end{pmatrix}.$$

The model is called Connes-Lott [25] model.

Application of noncommutative geometry to $SU(3) \times SU(2) \times U(1)$ symmetric standard model was done by Connes and Lott [25], and Chamseddine and Connes [45, 47]. Noncommutative geometry appears when one incorporates quaternion in analysis, since for quaternions p and q, results of multiplication $p \cdot q$ and $q \cdot p$ differes, if $p \neq q$. This additional degrees of freedom, or extra dimension was used to cover chiral-even and chiral-odd manifolds.

The Lagrangian they took consists of

- The pure gauge boson part $\mathcal{L}_G$.
- The fermion kinetic term $\mathcal{L}_f$.
- Kinetic terms for Higgs fields $\mathcal{L}_\phi$.

- The Yukawa couplingof Higgs fields with fermion $\mathcal{L}_Y$.
- The Higgs self-interaction $\mathcal{L}_V$.

For a compact Riemannian spin manifold M and a self-adjoint Dirac operator $D = \partial_M$ acting in the Hilbert space h of L^2 spinors on the manifold M, Connes et al constructed Yang-Mills action functional, expressed as a triple $(\mathcal{A}, h, D)$, where $\mathcal{A}$ is the von Neumann algebra on $\mathcal{M}$:

$$\mathcal{A} = C^\infty(M) \otimes \mathcal{A}_F$$

where $\mathcal{A}_F = \mathbf{C} + \mathbf{H} + M_3(\mathbf{C})$. Using a spectral geometry on $\mathcal{A}_F$ denoted as $(\mathcal{H}_F, D_F)$, the Hamiltonian and the Dirac operator are expressed as

$$\mathcal{H} = L^2(M, S) \otimes \mathcal{H}_F, \quad D = i\gamma^\mu \partial_\mu M \otimes 1 + \gamma_5 \otimes D_F$$

where ∂^μ_M means derivative on the manifold M.

They developed a theory of manifold in noncommutative geometry, using Hamiltonian algebra of quaternions $\mathbf{H}$ which is characterised by an element $x \in M_2(\mathbf{C})$ that satisfies by $g \in U(2)$ and J such that $JgJ^{-1} = \bar{g}$, the relation $JxJ^{-1} = \bar{x}$.

In order to reproduce $U(1) \times SU(2) \times SU(3)$ algebra $\mathcal{A} = \mathbf{C} \oplus \mathbf{H}$ and $\mathcal{B} = \mathbf{C} + M_3(\mathbf{C})$ were defined as $*$-algebra.

The Hilbert space is

$$h = h_0 \oplus (h_1 \otimes \mathbf{C}^3).$$

The $\mathcal{A}$ bi-module structure is given for $\lambda \in \mathbf{C}$ and $q \in \mathbf{H}$

$$\begin{aligned} (\lambda, q)(q_1, q_2) &= (\lambda q_1, q q_2) \\ (q_1, q_2)(\lambda, q) &= (q_1 q, q_2 \lambda) \end{aligned}$$

and

$$d(\lambda, q) = (q - \lambda, \lambda - q) \in \mathbf{H} \oplus \mathbf{H}, \quad (q_1, q_2)^* = (\bar{q}_1, \bar{q}_2)$$

for $q \in \mathbf{H}$. A pair of differential forms $(F, G) \in \Omega^2_D$ consists of i) $(2,0)$ type: $\mathbf{C}$ valued 2-form F, and $\mathbf{H}$ vaued 2-form G, ii) $(1,1)$ type due to quaternonic 1-forms (ω_1, ω_2), iii) $(0,2)$ type due to a pair (q_1, q_2). There is no ghosts.

The structure of the Clifford algebra of the finite space F given by the linear map from the space $\mathbf{H}\oplus\mathbf{H}$ of 1-forms into the algebra of 1-forms $\sum_i a_i db_i$ and $\sum_i a_i[D, b_i]$, where

$$D = \begin{bmatrix} 0 & \partial \\ \partial^* & 0 \end{bmatrix}$$

where $\partial = \delta_1 - i\delta_2$ and $\partial^* = -\delta_1 - i\delta_2$ have different chilarity, remained as a problem [45]. The problem of CP violation was discussed in [25] following the model of Peccei and Quinn [116, 117].

Orthogonal transformations in Euclidean E^n space, $O(n)$ denotes the real orthogonal transformation, which is the transitive group on the unit $(n-1)$ sphere $\mathbf{S}^{n-1}$. If $x_0 \in \mathbf{S}^{n-1}$ is fixed, subgroup leaving x_0 fixed is orthogonal group O^{n-1}, and we consider

$$\mathbf{S}^{n-1} = O(n)/O(n-1).$$

One observes for $n = 4$,

$$SO(4) \simeq \frac{\mathbf{S}^3 \times \mathbf{S}^3}{\{(1,1),(-1,-1)\}}$$

or $M_4(\mathbf{R}) \simeq \mathbf{H}\otimes\mathbf{H}$ [53].

In $SU(2)$ lattice QCD, a gauge transformation is written as

$${}^U A_\mu = U A_\mu U^\dagger + U\partial_\mu U^\dagger,$$

where $U \in SU(2)$ is a mapping from $\mathbf{R}^3$ to $\mathbf{S}^3$.

In the case of $SU(3)$, using complex variables we consider $GL(2,\mathbf{C})$ transformation. Constructions of noncommutative gravity theories from background space-times of noncommutative geometry are reviewed in [48].

5. Clifford Algebra on the Manifold of $\mathbf{S}^1 \times \mathbf{S}^3$

In Clifford algebra, finite dimensional division algebra $\mathbf{R}, \mathbf{C}$ and $\mathbf{H}$ can be treated equivalently. Quaternions satisfy the division axiom - that the product of two factprs cannot vanish without either factor vanishing. We consider the Lorentz transformations of a quaternion $q \to q^*$

$$q = \mathbf{I}q_0 + \mathbf{i}q_1 + \mathbf{j}q_2 + \mathbf{k}q_3$$

that satisfy $q_0^2 - q_1^2 - q_2^2 - q_3^2$ invariant. A transformation for complex numbers

$$z^* = \frac{az+b}{cz+d}$$

which is general linear transformation is too general.

Dirac has choosen [52] $q = uv^{-1}$, $u, v \in \mathbf{H}$ with any quaternion λ

$$q = u\lambda\lambda^{-1}v^{-1} = u\lambda(v\lambda)^{-1}.$$

To transform q to q^*, u, v are transformed as

$$u^* = au + bv, \quad v^* = cu + dv$$

where a, b, c, d are arbitrary quaternions, and $q^* = u^* v^{*-1}$

By the above transformations

$$\begin{aligned} q^* &= (au+bv)(cu+dv)^{-1} = (au+bv)v^{-1}v(cu+dv)^{-1} \\ &= (au+bv)v^{-1}[(cu+dv)v^{-1}]^{-1} = (aq+b)(cq+d)^{-1}. \end{aligned}$$

Alternatively when $q = \alpha^{-1}\beta$

$$q^* = (qa+b)^{-1}(qc+d).$$

The transformation of previous assignment $q = uv^{-1}$ gives

$$q^* = a(q + a^{-1}b)[c(q + c^{-1}d)]^{-1} = a(q + a^{-1}b)(q + c^{-1}d)^{-1}c^{-1}.$$

which yields a necessary condition $a^{-1}b \neq c^{-1}d$

The transformation $\tilde{q} = (a'q^* + b')(c'q^* + d')^{-1} = \tilde{u}\tilde{v}^{-1}$ and $\tilde{q} = q$ yields

$$\begin{aligned} &a'b + b'd = 0, \quad c'a + d'c = 0 \\ &a'a + b'c = c'b + d'd = m \end{aligned}$$

where m is a real number, not zero. The condition can be satisfied by [52]

$$\begin{aligned} a' &= \frac{m}{b^{-1}a + d^{-1}c} b^{-1} = \frac{m}{a - bd^{-1}c}, \\ b' &= \frac{-m}{b^{-1}a - d^{-1}c} d^{-1} = \frac{m}{c - db^{-1}a}, \\ c' &= \frac{m}{a^{-1}b - c^{-1}d} a^{-1} = \frac{m}{b - ac^{-1}d}, \\ d' &= \frac{-m}{a^{-1}b - c^{-1}d} c^{-1} = \frac{m}{d - ca^{-1}b}. \end{aligned}$$

In order to make q, q^* and $\tilde{q}$ satisfy Lorentz group relations,

$$Q_1 = u\bar{v}, \quad Q_2 = u\bar{u}, \quad Q_3 = v\bar{v}$$

are defined, and in Q_1, u, v are replaced by $u\lambda, v\lambda$, which yields

$$u\lambda(\bar{v\lambda}) = u\lambda\bar{\lambda}\bar{v} = Q_1\lambda\bar{\lambda}$$

Put $Q_1 = \mathbf{I}X_0 + \mathbf{i}X_1 + \mathbf{j}X_2 + \mathbf{k}X_3$, where $X_0, X_1, X_2, X_3 \in \mathbf{R}$, and

$$Q_2 = X_4 - X_5, \quad Q_3 = X_4 + X_5$$

three Qs define six real numbers $X_0, X_1, X_2, X_3, X_4, X_5$. If u, v are replaced by $u\lambda, v\lambda$, all X are multiplied by $\lambda\bar{\lambda}$ and their ratios are not changed.

From relations $Q_1\bar{Q}_1 = u\bar{v}v\bar{u} = uQ_3\bar{u} = Q_2Q_3$. Hence

$$X_0^2 + X_1^2 + X_2^2 + X_3^3 = X_4^2 - X_5^2.$$

By choosing $\tilde{q}$, one can define new X's and to make the transformations satisfy the Lorentz group, one can impose restictions of choosing planes $X_0 = 0, \quad X_5 = 0$, which is equivalent to

$$u\bar{v} + v\bar{u} = 0, \quad u\bar{u} + v\bar{v} = 0.$$

The transformation that satisfy Lorentz group is

$$q^* = (aq \pm \mu a)(-\mu\bar{a}q \pm a)^{-1}.$$

Four quantities $\xi_\nu = X_\nu/X_5$ and $\eta_\nu = X_\nu/X_0$ ($\nu = 1,2,3,4$) transform as vectors in space-time, and we choose ξ for analysis. A choice $u = q, v = 1$ gives

$$\xi_i = \frac{2q_i}{1-|q|^2}(i=1,2,3), \quad \xi_4 = \frac{1+|q|^2}{1-|q|^2}$$

and

$${\xi_4}^2 - {\xi_1}^2 - {\xi_2}^2 - {\xi_3}^2 = 1 + \frac{4q_0^2}{(1-|q|^2)^2}.$$

The chiral symmetry in complex projective space $\mathbf{CP}^{N-1}$ was studied also by d'Adda et al. [139, 140]. They considered spontaneous symmetry breaking of $SU(N)$ to $SU(N-1)\otimes U(1)$ due to instantons. Laglangian of N^2-1 scalar fields is defined as

$$\mathcal{L} = \frac{1}{2}Tr\partial_\mu\phi\partial^\mu\phi - \lambda Tr P(\phi)$$

where $P(\phi)$ is some polynomials [75]. Choosing N dimensional column vector of unit length $z^\dagger z = 1$, $\phi = g_0^{-1}[N^{1/2}zz^\dagger - N^{-1/2}I]$, where g_0 is a parameter derived from $P(\phi)$.

After proper normalization, and taking currents in the system as

$$j_\mu = \frac{1}{2i}[z^\dagger\partial_\mu z - (\partial_\mu z^\dagger)z]$$

the Lagrangian becomes

$$\mathcal{L} = \partial_\mu z^\dagger \partial^\mu z - g_0^2 N^{-1} j_\mu j^\mu$$

with the constraint $z^\dagger z = N/g_0^2$.

When $N = 4$, the dimension of the real Goldstone fields becomes

$$dimSU(4) - dimU(4-1) = 15 - (4-1)^2 = 6$$

which agrees with the dimension of X fields.

Kodaira showed in projective space $\mathbf{P}^3$, the cohomology groups of $\mathbf{P}^3$ with coefficients in $\Omega(\mathbf{C})$: $H^1(\mathbf{P}^3, \mathbf{C}) = 0$ [63] and one can define 2 algebraic curves on two dimensional surface. In projective space, zero-mode Dirac wave function contains a parameter that fixes the scale of the system.

Defining a compex manifold W as a domain $\mathbf{C}^2 - (0,0)$, and a surface S homeomorphic to $\mathbf{S}^1 \times \mathbf{S}^3$ and denoting Z_t an infinite cyclic group generated by a linear transformation $(z_1, z_2) \to (tz_1, tz_2)$ where $0 < |t| < 1$ and $S_t = W/Z_t$, and $\mathbf{S}^3$ is defined by $|z_1|^2 + |z_2|^2 = 1$. He prooved existence of local complex analytical coordinates (w_j, z_j) which define

$$\theta = \sum_{\alpha=1}^{3} \theta_j^\alpha \frac{\partial}{\partial z_j^\alpha}, \quad e^{cz_j} dw_j = \kappa^{m_j k} e^{cz_k} dw_k$$

on $U_j \cap U_k$, where κ is 0 at $w_k = 0$, and not zero elsewhere, c is a constant, and w_j is a holomorphic function of w_k independent of z_j,

$$w_j = f_{jk}(w_k)$$

on $U_j \cap U_k$.

Embedding $\mathbf{S}^1 \times \mathbf{S}^3$ on which quarterion $\mathbf{H}$ operates on $\mathbf{S}^3$ in $\mathbf{R}^4$ and a parameter for the Laplace transform operates on $\mathbf{S}^1$ can be compared with conventional embedding $\mathbf{S}^3 \times \mathbf{R}$ in $\mathbf{R}^4$, since local complex analytic coordinates (w_j, z_j) defined respectively on U_j such that the sheaf over S is denoted as Θ

Unitary transformation U^n operating in complex $n-$space is also transitive on the unit $(2n - 1)$ sphere, and we consider

$$\mathbf{S}^{2n-1} = U(n)/U(n-1)$$

and $n = 2$. Using quaternion variables, we can consider $U(1, \mathbf{H})$ as the quotient group [65].

The group $SU(2)$ is a $U(1)$-bundle over $\mathbf{S}^2$ and the projection of the bundle onto its base space is called the *Hopf* fibration [8]. We extend the theory to bundles over $\mathbf{S}^3 = O(4)/O(3)$. Hopf manifold is not an algebraic manifold, but the Lie group $GL(2, \mathbf{C})$ can operate on w_j [62].

On Hopf surface $W = \mathbf{C}^2 - \{0\}$, one defines automorphisms

$$g_t : (z_1, .z_2) \to (\alpha z_1 + tz_2, \alpha z_2), \quad 0 < |\alpha| < 1, t \in \mathbf{C}.$$

$G_t = \{g_t^m | m \in \mathbf{Z}\}$ is automorphic and do not have fixed points, since

$$g_t^m : (z_1, .z_2) \to (\alpha^m z_1 + m\alpha^{m-1} tz_2, \alpha^m z_2).$$

$M_t = \mathbf{C}/G_t$ is also a Hopf surface. Complex structure of M_0 and M_t, t not zero, are different. One dimensional algebraic manifold is called an algebraic curve. On $M_t, t \neq 0$, the regular linearly independent vector fields are

$c_1(z_1\frac{\partial}{\partial z_1} + z_2\frac{\partial}{\partial z_2})$ and $c_2 z_2 \frac{\partial}{\partial z_1}$ [64]. One can choose 2 algebraic curves $t = 0$ to $t = 1$ and $t = 1$ to $t = 0$ on the Hopf surface, which are related by the Holonomy group [57, 58]. When points A and C are on a sphere S, holonomy group $H_A(S)$ and $H_C(S)$ are isomorphic [58], but on a manifold $S^1 \times S^3$ $H_A(S)$ and $H_C(S)$ can have different parallel transformations. It is an interesting observation concerning the self-energy of electrons.

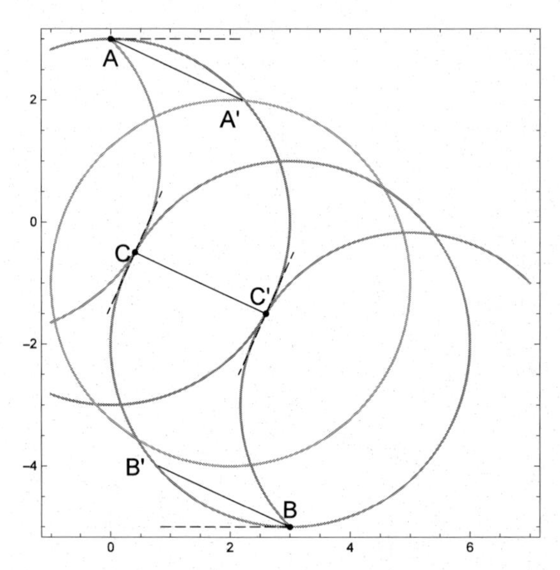

Figure 1. Holonomy curve $AC'B + BCA$, and $CBC' + C'AC$. The parallel transformations of the former curve have the direction parallel to the x axis, as shown by dashed lines from A and B. The parallel transformations of the latter curve have the direction parallel to CC', from A and B respectively, as AA' and BB'.

In the LFHQCD, one considers the light-front Lorentz transformation in evaluations of mass of a charged particle from its Poynting vector, where magnetic field plays an essential role. Magnetic field of QCD is not well known, but in QED, the magnetic field has hysteresis effect, and we expect it is same in QCD. The electro magnetic mass of a charged particle is measured in the rest frame, and the electric field plays an essential role. Due to the Coleman-Mandula theorem [108], time dependent selfmass cannot be Lorentz invariant, and the two different Poincaré stresses or difference of self-energy from static electrostatic energy and from dynamic poyiting vector supports the framework of the complex projective space $\mathbf{P}^3$ in analyses. As in LFHQCD [28], this framework does not fix the absolute scale of spectroscopy but fixes the ratios in the supersymmetric hadron spectroscopy.

DISCUSSION AND CONCLUSION

We reviewed properties of Yang-Mills action, which is described by functions $F' = g^{-1}Fg$, and Maxwell equation appears when $g = 1$. When spinors on manifolds are introduced and the gauge transformation is extended to $g = U(1) \times SU(2) \times U(3)$, the standard model of hadrons described by quarks and gluons was embedded in $\mathbf{R}^4$ space.

In QED, only transverse components of photons interact with leptons, and Gupta-Bleuler prescription [110] to cancel non-physical degrees of freedom was proposed. The solution of the $4/3$ problem by modifications of the Maxwell equation proposed by Dirac [3] was examined by Yaghjian [16]. He added time-dependent external force which is necessary for stability, and performed the Fourier-Laplace transformation to get spectra, but the structure of an electron could not be fixed.

In QCD, due to non-linearity of quark-gluon interactions, cancellation of non-physical degrees of freedom of gluons was performed by Faddeev-Popov ghosts. The Faddeev-Popov procedure contained the Gribov copy problem, which shows that the gauge fixing can not be done uniquely. Restricting gauge orbits to the fundamental modular region and models satisfying BRST symmetry was proposed, and by SU(3) lattice simulations suppression of the gluon propagator and enhancement of the ghost propagator in infrared region was confirmed.

The requirement of ghosts is 4-dimensional QCD is related to the treatment of chiral symmetry or γ^5 symmetry of Dirac fermions. Since exact BRST symmetry is difficult to verify, construction of renormalizable QCD theories without ghosts in the Hilbert space was tried by several researchers, by adding an extra dimension.

In light front holographic QCD (LFHQCD), Pauli-Lubanski relativistic spin approach was adopted and subtraction of non-physical gauge freedom outside proper light-cone was performed. In the approach of non-commutative geometry, covering of manifolds with positive chirarity and manifolds with negative chirality was performed using quaternion vectors.

If the divergent electromagnetic self energy is expressed as $\Delta(m^2)_e = m^2\frac{3e^2}{2\pi}log(\Lambda^2/m^2)$, where Λ is the cut off mass, and the photon propagator is $\frac{1}{q^2}(\frac{-\Lambda^2}{-\Lambda^2-q^2})$, electro magnetic mass difference of proton and neutron mass would become [134]

$$\Delta(M^2)_{proton} - \Delta(M^2)_{neutron} = -1.2934\text{MeV}.$$

Nucleons have spin 3/2 Δ resonance states, and extension of color $SU(3)_{color}$ to $SU(6)_{Color-LeptonNumber}$ was tried to explain hadron spectroscopy [135, 136] but the model $SU(6)_{Flavor-Spin}$ [29] was more successful. Jaffe [137] proposed combination of flavor $SU(3)$ (no charm contribution) and spin $SU(2)$ produces flavor singet stable dihyperon states. Incorpolation of quark and gluon degrees of freedom in deuteron form factor was done in [31, 33]. Prediction of tetraquark state spectroscopy in LFHQCD is given in [35]

In the Nahm-ADHM formalism, quaternion vectors on $\mathbf{S}^3 \times \mathbf{R}$, which is isomolphic to T^4, energy of zero momentum modes were investigated to fix the scale of the glueball mass [96, 102]. Effects of holonomy group on orbits on the $\mathbf{S}^1 \times \mathbf{S}^3$ and maximum conformality are to be investigated. There are arguments about mathematical observation in Hilbert space and physical semantics [142] If we observe mathematical objects in the projected space and physical semantics is correct, we solve the 4/3 problem.

I could not review experimental results which can be attributed to symmetry properties of gluons, like glue-balls , gluon jets etc which are shown by CMS, ATLAS, LEP and other groups. Theoretical topics were rather specified to algebra related to discrete topology. Spectroscopy of mesons and baryons due to quarks interacing through gluons related to the chiral symmetry were discussed. I think the new analysing method using complex projected spaces and $\mathbf{H}$ with an extra dimension, will give new insight on CP violation and hysteresis [143].

ACKNOWLEDMENT

I thank Prof. Stanley Brodsky for presentation of AdS/QCD in 2006 and continuous discussions; Dr. Marc Chemtob at PTI at Saclay for the instruction from 1973 to 1974 and critical reading of this article; Prof. Amand Faessler for instructing quark-gluon dynamics from 1981 to 1987; the late Prof. Konrad Bleuler for attracting my attention in 1989 to works of Prof. Alain Connes, when I was a research fellow of Prof. Max Huber at Bonn University; and the late Prof. Hideo Nakajima for collaborations on lattice QCD simulation from 1999 to 2007. Thanks are also due to Dr. Naoki Kondo at Teikyo University, Dr. Serge Dos Santos at INSA in Blois for communications, Libraries of Tokyo Institute of Technology and Libraries of the University of Tokyo for allowing consultation of references.

References

[1] Feynman, Richard P. , Leighton, Robert B. and Sands, Mattew.1964. *The Feynman Lectures on Physics II*, Addison-Wesley Pub. Reading, Massachusetts, Palo Alto, London.

[2] Poincaré, Henri 1905: *Translation On the Dynamics of the Electron*, https: //en. wikisource. org/wiki/On_the_Dynamics_of_the_Electron. (July)

[3] Dirac, P.A.M. 1938. "Classical theory of radiating electrons", *Proc. Roy. Soc. London, Ser.* **A167**, 148-168.

[4] Dyson, Freeman J. 1948. "The Radiation Theories of Tomonaga, Schwinger, and Feynman", *Phys. Rev.* **75** 486-502.

[5] Dyson, Freeman J. 1949. "The S-matrix in Quantum Electrodynamics", *Phys. Rev.* **75** 1736-1755.

[6] Schwinger, Julian 1949. "On the Classical Radiation of Accelerated Electrons", *Phys. Rev.* **75** (12) 1912-1925.

[7] Creutz, Michael 1983. *Quarks, Gluons and Lattices*. Cambridge University Press.

[8] Madore, John 1999. *An Introduction to Noncommutative Differential Geometry and its Physical Applications* 2nd Edition, Cambridge University Press.

[9] Dirac, P.A.M. 1931. “Quantized Singularities in the Electromagnetic Field“, *Proc. Roy. Soc. (London)* **A 133**, 60.

[10] Ebert, Dietmer 1989. *Eichtheorien; Grundlage der Elementarteilchenphysik* [*Gauge theories; Basis of elementary particle physics*], VCH, Weinheim.

[11] Schwinger, Julian 1983. “Self mass of an electron revisited”, *Found. Phys.* **13** 373

[12] Jackson, J.D. 1999. *Classical Electrodynamics*, 3rd Ed. John Wiley and Sons Inc, New York; Translated to Japanese by Nishida, Minoru, 2003. Yoshioka Shoten, Kyoto.

[13] Rohrlich, F. “The dynamics of a chrged sphere and the electron”, *Am. J. Phys.* **65** (11) 1051-

[14] Rohrlich, F. *Classical Charged Particles*, 3rd Ed., World Scientific Pub. New York.

[15] Medina, Rodorigo 2006. “Radiationreaction of a classical quasi-rigid extended particle, *J. Phys A: Math Gen.* **39** 3801.

[16] Yaghjian, A. 2006. *Relativistic Dynamics of Charged Sphere: Updating the Lorentz-Abragham Model* Springer, New York.

[17] Englert, F. and Brout, R. 1964. “Broken Symmetry and the Mass of Gauge Vector Mosons“, *Phys. Rev. Lett.* **13** 321.

[18] Higgs, Peter W. 1966. “ Spontaneous Symmetry Breakdown without Massless Bosons“, *Phys. Rev.* **145** 1156-1163. Phys. Lett. **12** 132; Phys. Rev. Lett. **13**, 508.

[19] Higgs, Peter W. 2014. “NovelLecture: Evading the Goldstone theorem”, *Review of Modern Physics* **86** 851-853.

[20] Kibble, T.W.B. 1967. “Symmetry Breaking in Non-Abelian Gauge Theories“, *Phys. Rev.* **155** 1554.

[21] Cheng, Ta-Pei and Li,Ling-Fong 1984, *Gauge theory of elementary particle physics*, Clarendon Press,

[22] ‘t Hooft, Gerard 1974. “Magnetic Monopoles in Unified Gauge Theories”, *Nucl. Phys.* **B 79**, 276-284.

[23] Polyakov, A.M. 1974. “Particle spectrum in quantum field theory“ *JETP Lett.* **20**, 194.

[24] Polyakov, A.M. 1987. *Gauge Fields and Strings, Contemporary Concepts in Physics*, Vol.3, Harwood Academic publishers, Chur Switzerland.

[25] Connes, Alain and Lott, John 1990. “Particle Models and Noncommutative Geometry”, *Nucl. Phys.* **B** (Proc.Supp.) 18B 29-47.

[26] Brodsky, Stanley J., Mojaza, Matin and Wu, Xing-Gang 2014. “Systematic Scale Setting to All Orders: The Principle of Maximum Conformality and Commensurate Scale Relations”, *Phys. Rev.* **D89** 014027

[27] de Téramond, Dosch, Hans G. and Brodsky, Stanley J. 2015, “Baryon Spectrum from Superconformal Quantum Mechanics and its Light-Front Holographic Embedding”, *Phys. Rev.* **D91** 045040; arXiv:1411.5243v2 [hep-ph]

[28] Dosch, Hans G., de Téramond, Guy F. and Brodsky, Stanley, J. 2015. “Superconformal Baryon-Meson Symmetry and Light-Front Holographic QCD”, *Phys. Rev.* **D97** 085016; arXiv:1501.00959v2[hep-th]

[29] Brodsky, Stanley J. 2007. “The Conformal Template and New Perspective for Quantum Chromodynamics”, *Progress of Theoretical Physics Supplement* **167** 76-94.

[30] Bashkanov, M., Brodsky, S.J., and Clement, H. 2013. “Novel Six-Quark Hidden-Color Dibaryon States in QCD”, *Phys. Lett.* **B727** 438-442, arXiv:1308.6404 [hep-ph].

[31] Brodsky, Stanley J., Ji, Chueng-Ryong, and Lepage, G.P. 1983. “Quantum Chromodynamic Prediction for the Deuteron Form Factor, *Phys. Rev. Lett.* **51** 83-86.

[32] Chiu, Kelly Y-Ju and Brodsky, Stanley J. 2017. "Angular momenrum conservation law in light-front quantum field theory", *Phys. Rev.* **D 95**, 065035.

[33] Brodsky, Stanley J., Chiu, Kelly Yu-Ju, Lansberg, Jean-Philippe and Yamanaka, Nodoka. 2019. *The gluon and charm content of the deuteron*, arXiv:1805.03173 v1 [hep-ph].

[34] Brodsky, Stanley j. 2018, "Supersymmetric and Conformal Features of Hadron Physics", *Universe* **2018**,4,1 20 1-9 (MDPI)

[35] Brodsky, Stanley J. 2019. "Color Confinement and Supersymmetric Properties of Hadron Physics from Light-Front Holography", *IOP Conf. Series: Journal of Physics: Conf. Series* **1137** 012027

[36] de Alfaro, Y. , Fubini, S. and Furlan, G. 1976. "Conformal invariance in quantum mechanics", *Nuovo Cim.* **A 34**, 569.

[37] Fubini, S. and Ravinovich, E. 1984. "Superconformal Quantum Mechanics", *Nucl. Phys.* **B245**, 17-24.

[38] Labelle, Patrick. 2010. *Supersymmetry DeMystified*, Mc Graw Hill. Com. New York.

[39] Chemtob, Marc 1984. "Instantons and Chiral Symmetry in Finite Temperature and Density QCD", *Physica Scripta* **29** (1) 17-36.

[40] Chemtob, Marc 1985. " The three flavor Skyrme model applied to the baryon octet and decimet", Nuovo Cim. **89A** 381.

[41] Chemtob, Marc 1987. "Skyrmion-Baryon Phenomenology in the Effective Gauged Chiral-SU(2) Action Approach", *Nucl. Phys.* **A466** 509-559.

[42] Witten, E. 1983. "Global aspects of current algebra", *Nucl. Phys.* **B223** 422-432.

[43] Witten, E. 1983. "Current algebra, baryons, and quark confinement", *Nucl. Phys.* **B223** 433-444.

[44] Connes, Alain 1990. *Géométrie non commutative* Inter-Editions, Paris, Translated to Japanese by Maruyama,Fumitsuna, Iwanami Shoten Pub. Tokyo.

[45] Connes, Alain 1994, *Noncommutative Geometry*, Academic Press, An Imprint of Elsevier, San Diego, New York, Boston, London, Sydney, Tokyo, Toronto.

[46] Connes, Alain 1995. "Noncommutative geometry and reality", *J. Math. Phys.* **36** (11) 6194-6231.

[47] Chamseddine, Ali H. and Connes, Alain 1997. "The Spectral Action Principle", *Commun. Math. Phys.* **186** 731-750.

[48] Monolakos, George, Manousselis, Pantelis and Zoupanos, George 2019. "Gauge Theories: From Kaluza-Klein to noncommutative gravity theories", Symmetry **2019**, 11, 856.

[49] Wess, Julius and Zumino, Bruno 1974. "Supergauge Transformations in four Dimensions", *Nucl. Phys.* **B70** 39-50.

[50] Wess, Julius and Zumino, Bruno 1990. "Covariant Differential Calculus on the Quantum Hyperplane", *Nucl Phys. B* (Proc suppl.) **18B** 302-312.

[51] Souriau, J. -M. 1970. *structures des systèmes dynamiques*, DUNOD, Paris.

[52] Dirac, P.A.M. 1945. "Application of Quaternions to Lorentz Transformations", *Proc. Roy. Irish Acad. (Dublin)*, **A 50** 261-270.

[53] Lounesto, Pertti 2001. *Clifford Algebras and Spinors* 2nd Edition, Cambridge University Press.

[54] Bott, Raoult and Tu, Loring W. 1982 , *Differential Forms in Algebraic Topology*, Springer-Verlag, New York Heidelberg Berlin.Academic Press,

[55] Garling, D. J. H. 2011. *Clifford Algebras: An Introduction*, London Mathematical Society, Student Texts 78, Cambridge University Press.

[56] Morais, Joao Pedro, Georgiev, Svetlin and Sproessig, Wolfgang 2014. *Real Quaternionic Calculus Handbook*, Birkhaeuser, Springer, Basel.

[57] Murakami, Shingo 1969. *Manifolds* (In Japanese) Kyoritsu Shuppan Pub. Tokyo.

[58] Carmo, Manfred. P. do 1976.*Differentialgeometrie von Kurven und Flaechen*, Friedr. Vieweg & Sohn, Braunschweig/Wiesbaden. p. 231 (Translated from the English original version by Grueter, M. 1983).

[59] Nash, Charles and Sen, Siddhartha 1983. *Topology and Geometry for Physicists* Dover Publications Inc. Mineola, New York.

[60] Morita, Katsusada 2011. *Quaternions, Octonions and the Dirac Theory*, (In Japanese), Nihon-Hyoronsha, Tokyo.

[61] Albererio, Sergio, Jost, J'urgen, Paycha, Sylvie and Scarlatti, Sergio 1997. *A Mathematical Introduction to String Theory*, Variational problems, geometric and probabilistic methods, Cambridge Univ. Press., Cambridge.

[62] Kodaira, Kunihiko 1964. "On the Structure of Compact Complex Analytic Surfaces I", *Amer. J. Math.* **86** 751-798.

[63] Kodaira, Kunihiko 1965. "Complex Structures on $S^1 \times S^3$", *Mathematics* **55** 240-243, Proc. N.A.S.

[64] Kodaira, Kunihiko 1992. *Theory of Complex Manifolds* (In Japanese), Iwanami Shoten Pub.

[65] Chevalley, Claude 1946. *Theory of Lie Groups I*, Princeton University Press; Asian text edition, 1965, Overseas Publications LTD. (Kaigai Shuppan Boeki K.K.), Tokyo.

[66] Steenrod, Norman 1951. *The Topology of fibre Bundles*, Princeton University Press.

[67] Porteous, Ian R. 1995. *Clifford Algebra and the Classical Groups*, Cambridge University Press.

[68] Becher, Peter, Boehm, Manfred and Joos, Hans 1981, *Eichtheorien der starken und electromagnetischen Wechselwirkung* [*Gauging theories of strong and electromagnetic interaction*], B.G. Teubner, Stuttgart.

[69] Becchi Carlo M. and Ridolfi, Giovanni 2006, *An Introduction to Relativitic Processes and the Standard model of Electroweak Interactions*, Springer-Verlag Italia.

[70] Faddeev, L.D. and Slavnov, A.A. 1990. *Gauge Fields -Introduction to Quantum Theory-*, Translated from the Russian Edition by Pontecorvo, G.B., Addison Wesley Publishing Company, Redwood City, California.

[71] Henneaux, Marc and Teitelboim, Claudio. 1992. *Quantization of Gauge Systems*, Princeton University Press.

[72] Faddeev, L. D. and Popov, V. N. 1967. "Feynman Diagrams for the Yang-Mills Field", *Phys. Lett.* **25B** 29-30.

[73] Gribov, V.N. 1978, "Quantization of Non-Abelian Gauge Theories", *Nucl. Phys.* **B138** 1-19.

[74] Fujikawa, Kazuo, 1979. "Comment on the Covariant Path Integral formalism in the Presence of Gribov Ambiguities", *Prog. Theor. Phys.* **61** 627-632.

[75] Coleman, Sidney 1985. *Aspects of Symmetry*, Selected Erice lectures, Cambridge University Press, Cambridge.

[76] Maskawa, Toshihide and Nakajima, Hideo, 1978. "How Dense Are the Coulomb Gauge Fixing Degeneracies? " *Prog. Theor. Phys.* **60** 1526.

[77] Kugo, Taichiro and Ojima, Izumi 1978. "Manifestly Covariant Cannonical Formulation of Yang-Mills Theories, Physical State subsidiary Conditions and Physical S-matrix Unitarity", *Phys. Lett.* **73B** 459-462.

[78] Kugo, Taichiro and Ojima, Izumi 1979. "Local Covariant Operator Formalism of Non-Abelian Gauge Theories and Quark Confinement Problem", *Prog. Theor. Phys. Supp.* **66** 1-130. *Errata: 1984 Prog. Theor, Phys.* **71** 1121

[79] Kugo, Taichiro 2000. “Color Confinement and Quartet Mechanism”, *Confinement 2000*. Ed. by Suganuma, H. et al., p. 51-59, World Scientific. Singapore

[80] Nakajima, Hideo and Furui, Sadataka, 2000. ”Numerical Test of the Kugo-Ojima Color Confinement Criterion”, *Confinement 2000*. Ed. by Suganuma, H. et al., p. 60-69, World Scientific. Singapore

[81] Nakajima, Hideo and Furui, Sadataka 2004. “Infrared Features of the Lattice Landau Gauge QCD”, *Nucl Phys* **B** (Proc Suppl.) 129&130 730-732.

[82] Furui, Sadataka and Nakajima, Hideo 2006. “Infrared features of unquenched Lattice Landau Gauge QCD”, *Few Body Systems* **40** 101-128 ; arXiv:0503029[hep-lat].

[83] Furui, Sadataka and Nakajima, Hideo 2006. “Effects of quark field on the ghost propagator of Lattice Landau Gauge QCD”, *Phys Rev.* **D73**, 094506; arXiv: 0602027[hep-lat].

[84] Beccchi, C., Rouet, A. and Stora, A. 1975. “Renormalization of the abelian higgs-Kibble model“, *Comm Math. Phys.* **42** 127.

[85] Zwanziger, Daniel. 1990. “Quantization of Gauge Fields, Clssical Gauge Invariance and Gluon Confinement”, *Nucl. Phys* **B345** 461-471.

[86] Zwanziger, Daniel. 1991. “Vanishing of Zero-momentum Lattice Gluon Propagator and Color Confinement”, *Nucl. Phys.* **B364** 127-161.

[87] Dashen, Roger, Hasslacher, Brosl and Neveu, André 1974. “Nonperturbative methods and extended hadron models in field theory. III. Four-dimensional non-Abelian models”, *Phys. Rev.* **D10** (12) 4138-4142.

[88] Yaffe, Laurece G. 1989. “Static solutions of SU(2) Higgs theory”, *Phys. Rev.* **D40** (10) 3463-3473.

[89] Milnor,J. 1963. “Morse Theory”, *Ann. of Math. Studies* 51, Princeton Univ. Press.; Translated to Japanese by Shiga, Kooji (1968) Yoshioka Shoten Pub.

[90] Braam, Prter J. and van Baal, Pierre 1989. "Nahm's transformaion for Instantons", *Commun. Math. Phys.* 267.

[91] Atiyah, M.F., Hitchin, N.J. , Drinfeld, V. G. and Manin, Yu. I. 1978. "Construction of Instantons". *Phys. Lett.* **65A** 185-187.

[92] van Baal, Pierre. 1992. " More (Thought on) Gribov Copies", *Nucl. Phys,* **B369** 259.

[93] van Baal, Pierre and Cutkosky, R.E. 1993. "Non-Perturbative Analysis, Gribov Horizons and the boundary of the Fundamental Domain", *Int. J. Mod. Phys. A* (Proc. Suppl.) 3A 323; 21st Conference on Differential Geometric Methods in Theoretical Physics (XXI DGM 1992)

[94] van Baal, Pierre and Hari Dass, N.D. 1992. "The theta dependence beyond steepest descent", *Nucl. Phys.* **B385** 185-226.

[95] van Baal, Pierre and van den Heuvel. 1994. "Zooming-in on the SU(2) fundamental domain", *Nucl. Phys.* **B417** 215-237.

[96] van Baal, Pierre. 1995. "Global issues in gauge fixing", arXiv: hep-th/9511119v1. Talk at the *ECT workshop "Non-perturbative approaches to QCD"*, Trento Italy.

[97] Zwanziger, Daniel. 1994. "Fundamental modular region, Boltzmann factor and area law in lattice theory", *Nucl. Phys.* **B412** 657-730.

[98] Fujikawa, Kazuo 1995, *BRST Symmetric Formulation of a Theory with Gribov-type Copies*, arXiv: hep-th/9510111.

[99] Alkofer, R. , Huber, M.Q. and Schwenzer, K. 2018. "Algorithmic derivation of Dyson-Schwinger equations", *Computer Physics Communications* **180** 965-976.

[100] Huber, M.Q., Alkofer, Reinhard and Sorella, Silbio P. 2010. "Non-perturbative analysis of the Gribov-Zwanziger action", arXiv: 1010.4802 v1[hep-th], Talk at the *Quark Confinement and the Hadron Spectrum IX Conference*, 2010.

[101] Watson, Peter and Alkofer, Reinhard. 2018. *Verifying the Kugo-Ojima Confinement Criterion in Landau Gauge Yang-Mills theory*, arXiv: hep-ph/ 0102332v2

[102] Pérez, Margarita Garcia and van Baal, Pierre 1994. "Sphalerons and other saddles from cooling" , *Nucl. Phys.* **B429** 451-473.

[103] Golterman, Maarten F.L. and Shamir, Yigal 1996. "A Gauge-fixing action for lattice gauge theories", *Phys. Lett.* **B399**, 148-155, arXiv: hep-lat/9608116 v2.

[104] Baulieu, Laurent and Zwanziger, Daniel 2001.*Bulk Quantization of Gauge Theories: Confined and Higgs Phases*, arXiv:hep-th/0107074.

[105] Zwanziger, Daniel 2003. *Non-perturbative Faddeev-Popov formula and infrared limit of QCD*, arXiv: hep-ph/0303028

[106] Schaden, Martin and Zwanziger, Daniel 2015. "BRST cohomology and physical space of the GZ model", *Phys. Rev.* **D92** 025001 ; arXiv: 1412.4823v3 [hep-ph]

[107] Schaden, Martin and Zwanziger, Daniel 2015. *Poincaré Symmetry of the GZ model*; arXiv: 1501.05974v2 [hep-th]

[108] Coleman, Sidney and Mandula, Jeffery 1967. "All Possible Symmetries of S Matrix", *Phys. Rev.* **159**(5) 1251.

[109] Haag, Rudolf, Sohnius, Martin and Lopuszanski, Jan T. 1975. "All possible generators of supersymmetries of the S-matrix", *Nucl. Phys.* **B88** 257-274.

[110] Bleuler, Konrad 1950, "Eine neue Methode zur Berechnung der longitudinalen und skalaren Photonen", *Helvetica Physica Acta* **23** 567-586.

[111] Nakajima, Hideo and Furui, Sadataka 1999. "The Landau gauge lattice QCD simulation and the gluon propagator", *Nucl. Phys.* **B** (Proc Suppl.) **73** 635-637.

[112] Nakajima, Hideo and Furui, Sadataka 2001. “A test of the Kugo-Ojima confinement criterion by lattice Landau gauge QCD simulations”, *Nucl. Phys.* **A680** 151c-154c.

[113] Furui, Sadataka and Nakajima, Hideo. 2003 *Infrared Features of the Landau Gauge QCD* arXiv:hep-lat/0305010.

[114] Furui, Sadataka and Nakajima, Hideo. 2004 “What the Gribov copy tells on the confinement and the theory of dynamical chiral symmetry breaking” *Phys. Rev.* **D70**, 094504(2004), arXiv:hep-lat/0403021.

[115] Bernard, Claud (MILC Collaboration) 2002. *Flavour Symmetry Breaking*, hep-lat/0111051v3.

[116] Peccei, R.D. and Quinn, Helen R. 1977, “Constraints imposed by CP conservation in the presence of pseudoparticles”, *Phys. Rev.* **D16** (6) 1791-1797.

[117] Weinberg, Steven 1978, “A New Light Boson? , *Phys. Rev.* Lett. **40** 223-225.

[118] Wilczek, F. “Problem of Strong P and T Invariance in the Presence of Instantons”, *Phys. Rev.* Lett. **40** (5) 279-282.

[119] Dunajski, Maciej 2010. *Solitons, Instantons and Twistors*, Oxford University Press.

[120] Pérez, Margarita G., Gpnzález-Arroyo, Antonio and Sastre, Alfonso 2009. “Adjoint fermion zero-modes for $SU(N)$ carolons”, *JHEP*; arXiv:0905.0645 v1 [hep-th]

[121] Mackey, George W. 1968. *Induced Representation of Groups and Quantum Mechanics*, W.A. Benjamin INC, New York, Amsterdam and Editore Boringhieri, Torino.

[122] ‘t Hooft, G. and Veltman, M. 1972. “Regularization and Renormalization of Gauge Fields”, *Nucl. Phys.* **B44** 189-213.

[123] ‘t Hooft, G. and Veltman, M. 1972. “Combinatorics of Gauge Fields”, *Nucl. Phys.* **B50** 318-353.

[124] Arafune, J. , Freund, P. G. O. and Goebel, C.J. 1975. "Topology of Higgs fields", *J. Math. Phys.* **16** 433-437.

[125] 't Hooft, Gerard 1976. "Symmetry Breaking through Bell-Jackiw Anomalies", *Phys. Rev.* Lett. **37** 8.

[126] 't Hooft, Gerard 2007. "Models for Confinement", *Progress of Theoretical Physics Supplement* **167** 144-154.

[127] Ji, Xiangdong 1998. "Lorentz symmetry and internal structure of the nucleon", *Phys. Rev.* **D58** 056003.

[128] Ji, Xiangdong 2006. "A Unified Picture for Single Transverse-Spin Asymmetries in Hard Processes", *Phys. Rev.* Lett. **97** 082002, arXiv: hep-ph/0602239 v1.

[129] Penrose, R. and Rindler, W. 1984. *Spinors and space-time*, vol 1, Cambridge Monographs on Mathematical Physics.

[130] Penrose, R. and Rindler, W. 1986. *Spinors and space-time*, vol 2, Cambridge Monographs on Mathematical Physics.

[131] Mardacena, Juan M. 1999. "The Large-N limit of super-conformal field theories and super gravity", *Int. J. Theor. Phys.* **38** 1113.

[132] Susskind, Leonard and Lindsey, James 2005. *An Introduction to Black Holes, Information and the String Theory Revolution, The Holographic Universe*, World Scientific, Singapore.

[133] Takeuchi, Gaishi. 1983. *Lie Algebra and Elementary Particle Physics*, (In Japanese) Shokabo Pub.

[134] Feynman, R. P. 1972. "Photon-Hadron Interactions", *Frontiers in Physics, Lecture note series*, W.A. Benjamin, Inc. Massachusetts.

[135] Dyson, Freeman J. and Xuong, Nguyen-Huu. 1964. "Y=2 States in SU(6) Theory" *Phys. Rev.* Lett. **13** 815-818, (Errata) 1965. *Phys. Rev.* Lett. **14** 339.

[136] Dyson, Freeman 2018. *Maker of Patterns, An Autobiography through Letters*, Liverright Publishing Corporation, New York. p. 312

[137] Jaffe, R. L. 1977. "Perhaps a Stable Dihyperon" *Phys Rev. Lett.* **38** 195-198, (Errata) 617.

[138] D' Adda, A, Di Vecchia, P. 1978. "Supersymmetry and Instantons", *Phys. Lett.* **73B** 162-166.

[139] D' Adda, A. , Luescher, M. and Di Vecchia, P. 1978. "A 1/n Expandable Series of Non-linear σ Models with Instantons", *Nucl. Phys.* **B146** 63-76.

[140] D' Adda, A, Di Vecchia, P , and Luescher, M. 1979. "Confinement and Chiral Symmetry Breaking in $\mathbf{CP}^{n-1}$ Models with Quarks", *Nucl. Phys.* **B153** 125-144.

[141] Heitler, W. 1954, *The Quantum Theory of Radiation* Oxford University Press, Oxford, 3rd Edition, (Translsted to Japanese by Sawada, Katsuro 1958. Yoshioka Shoten Pub. Kyoto.), Appendix 6 for Chapters 25 and 26, Appendix 7 for Chapters 4 and 29.

[142] Weiszsaecker, C. -F. Frhr. v. 1987. "Die philosophische Interpretation der modernenPhysik" *Nova Acta Leopoldina*, **207**, 37/2, Johann Ambrosius Barth, Leibzig.

[143] Furui, Sadataka and dos Santos, Serge 2018. *Theoretical Syudy of Memristor and Time Reversal based Nonlinear Elastic Wave Spectroscopy*, submitted to MDPI.

In: Horizons in World Physics. Volume 302 ISBN: 978-1-53617-180-8
Editor: Albert Reimer

Chapter 7

GLUONS AT SMALL *X*

Akbari Jahan*

Department of Physics, North Eastern Regional Institute of Science and Technology, Nirjuli, Arunachal Pradesh, India

ABSTRACT

Gluons are vector gauge bosons that mediate strong interactions of quarks in quantum chromodynamics (QCD). The gluon distribution is notoriously difficult to measure over the whole Bjorken-x range. In the small x region ($x \leq 0.1$), the gluon distribution is reasonably well measured by the HERA experiments. This does not help at larger x because the scaling violations due to quarks radiating gluons dominate over the gluon splitting contribution. A very strong constraint for the gluon distribution at medium and large x is, however, given by the energy momentum sum rule. At sufficiently small x, non-linear gluon interaction effects have been considered in order to moderate the rise of the cross section. The measurements of the structure functions by deep inelastic scattering (DIS) processes in the small x region have opened up a new era in parton density measurements inside hadrons. The structure functions

* Corresponding Author's E-mail: akbari.jahan@gmail.com.

reflect the momentum distribution of partons in a nucleon. Their steep increase towards small x, observed at HERA, also indicates a similar increase in gluon distribution towards small x in perturbative QCD. Therefore, gluon distribution is the observable that governs the physics of high-energy processes in QCD. Despite the difficulty to study their behavior, we shall try to have a closer look at gluons in this chapter.

1. INTRODUCTION

One of the main theoretical challenges facing physicists is to understand how the tiny elementary particles give rise to most of the mass in the visible universe. These tiny particles, called quarks and gluons, are the building blocks for bigger particles such as protons and neutrons, which in turn form the larger particles, i.e., atoms. However, quarks and gluons behave way differently from those of the bigger particles, making them very difficult to study. Gluons are what hold the quarks together to make the larger particles (see Figure 1). They act as the exchange particle for the strong force between the quarks, and are analogous with the exchange of photons in the electromagnetic interactions between two charged particles.

Gluons are difficult to study because although they exist in nature all the time, they are very small and require very high energy to break them away from quarks. Physicists have, however, been able to learn more about them from experiments at Hadron Electron Ring Accelerator (HERA), Hamburg and the Large Hadron Collider (LHC) at CERN, Geneva.

Technically, gluons are vector gauge bosons having spin 1 and negative intrinsic parity and they mediate the strong interactions of quarks. The range of the strong force is limited by the fact that the gluons interact with each other as well as with quarks in the context of quark confinement. These properties contrast them with photons, which are massless and of infinite range. Since gluons carry color charge, they can interact with other gluons too. Simple schematic representations of the 3-gluon and 4-gluon vertices are shown in Figure 2.

Figure 1. Gluons (the wavy lines) holding the quarks (u, u and d) together in proton.

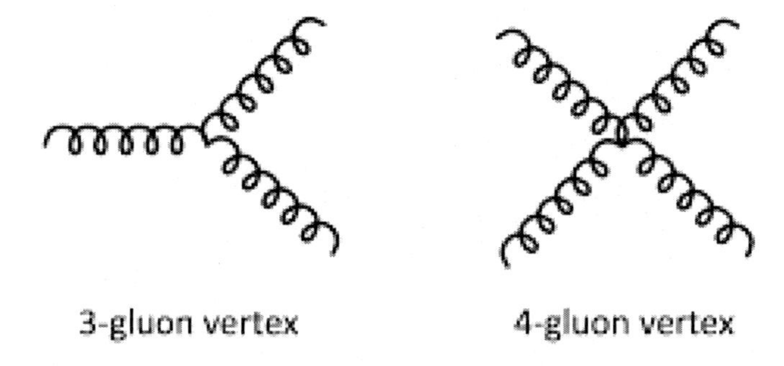

Figure 2. Self coupling of gluons.

Until the nineteen sixties, it was widely believed that particles like proton and neutron are as elementary as electron or muon. But in 1964, Gell-Mann [1] and Zweig [2] independently proposed a model in which the elementary particles are the so-called quarks, out of which hadrons like protons and neutrons are composed. In the late nineteen sixties, high-energy electron-proton scattering at the Stanford Linear Accelerator Center (SLAC) provided the first evidence for substructure within the proton [3-5]. Today, it is a well-established fact that nucleons (protons and neutrons collectively) are composed of quarks and gluons, whose interactions are described by the theory of Quantum Chromodynamics (QCD). QCD describes a wealth of physical phenomena, from the structure of nuclei to the cross sections of the highest energy elementary particle collisions. We outline it briefly in the subsequent section.

1.1. Theoretical Framework

1.1.1. Deep Inelastic Scattering and Quark Parton Model

Deep inelastic lepton-nucleon scattering (DIS) plays a key role in determining the partonic structure of the proton. The DIS experiments have revealed the structure of hadrons being made out of constituents and the interaction between the later. There have been two main strands of interest in experiments on DIS since the initial observation of Bjorken scaling. Firstly, they are used to investigate the theory of the strong interaction and secondly, they are used to determine the momentum distributions of partons within the nucleon [6]. The DIS process at HERA is characterized by the exchange of a virtual gauge boson between the interacting lepton and proton.

On the other hand, the quark-parton model (QPM) grew out of the attempt by Feynman [7] to provide a simple physical picture of the scaling that had been predicted by Bjorken [8] and observed in the first high-energy deep inelastic electron scattering experiments at SLAC [3-5] where the structure function $F_2(x,Q^2)$ was observed to be independent of Q^2 for x values around $x \sim 0.3$. The model states that the nucleon is full of point-like non-interacting scattering centers known as partons. The model expresses the hadron structure functions in terms of parton distribution functions (PDFs) that give the longitudinal momentum distribution of the partons in the given hadron. The PDFs are found from experimental data in a given process. J. D. Bjorken predicted that the hadronic factor in the cross section would depend only on the ratio $x = (-q^2)/(2P.q) = (-q^2)/(2M\nu)$, rather than on ν and $-q^2$ separately, on the basis of an algebra of local currents [8]. This property, called *scaling*, was expected to hold in the "deep inelastic" limit in which the energy transfer and momentum transfer are much larger than the target hadron mass. Feynman interpreted scaling in terms of constituents of the nucleon that he called "partons" [7]. Bjorken and Paschos [9, 10] gave early discussions of electron-nucleon and neutrino-nucleon scattering in the deep inelastic limit. The Bjorken-x can be identified with the fraction of the longitudinal hadron momentum carried by a given parton.

Feynman gave the relation between the structure function of proton and the quark densities. Denoting $q_i(x)$ as parton densities for quarks and $\overline{q_i}(x)$ for the corresponding antiquarks, the structure function F_2 is expressed in terms of quark densities as

$$F_2(x) = \sum_i e_i^2 \, x \, \{q_i(x) + \overline{q_i}(x)\} \tag{1}$$

This relation describes the probability distribution to find within the proton a parton specie i with momentum fraction x of the proton momentum and with electric charge e_i.

The model predicts that the integral over all momenta, carried by charged partons within the proton, is equal to one, i.e.,

$$\sum_i \int_0^1 x\left(q_i(x) + \overline{q}_i(x)\right) dx = 1 \tag{2}$$

where the summation is over all quark species in the proton. In the late nineteen sixties, experiments had shown that quarks carry only about 50% of the proton momenta and thus another parton specie, electrically neutral, must exist within the proton, subsequently identified with gluons. This gave impetus to the development of the theory of QCD.

1.1.2. Quantum Chromodynamics

Quantum Chromodynamics (QCD) is a remarkably successful and rich theory that describes strong interactions of quarks via intermediate massless vector bosons called gluons. In other words, QCD is a non-Abelian gauge theory of the strong interactions between quarks and gluons, which allows us to reconcile short distance freedom with long distance confinement. This comes about because the strength of strong interaction is variable. Quarks carry a charge called *color* and gluons carry combinations of colors. Such color charge of gluons enables gluons to couple to themselves.

One of the most interesting features of QCD is the property of the asymptotic freedom: the strong coupling constant (α_s) is small at large momenta (short distances), and it is large at small momenta (large distances) [11, 12]. Finding the scale that determines the value of the characteristic running QCD coupling is one of the central questions for high-energy scattering physics, important for the theoretical description of both the hadronic and nuclear scattering processes.

1.2. Measurement of Gluon Distribution at HERA

The gluon distribution is notoriously difficult to measure over the whole x range. In the small x region ($x \leq 0.1$), the gluon distribution is reasonably well measured by the HERA experiments. This does not help at larger x because there the scaling violations due to quarks radiating gluons dominate over the gluon splitting contribution. A very strong constraint for the gluon distribution at medium and large x is, however, given by the energy momentum sum rule. The number and momentum distribution of the gluons in proton, i.e., the gluon density, have been measured by two experiments, H1 and ZEUS during the years 1996-2007. The gluon contribution to the proton spin has been studied by the HERMES experiment at HERA [13].

The Q^2 evolution of the proton structure function $F_2(x,Q^2)$ is related to the gluon momentum distribution in the proton, $G(x,Q^2)$ and the strong interaction coupling constant (α_s). The strong scaling violations observed at small x are attributed to the high gluon density in the proton. At sufficiently small x, non-linear gluon interaction effects have been considered in order to moderate the rise of the cross-section in accordance with unitarity requirements [14]. $F_2(x,Q^2)$, which contains both valence and sea quark distributions, will rise with Q^2 at small x, where sea quarks dominate, and fall with Q^2 at large x, where valence quarks dominate.

The structure function reveals several significant information of the nucleon. If the proton is comprised of only one quark, it carries the entire momentum of the proton.

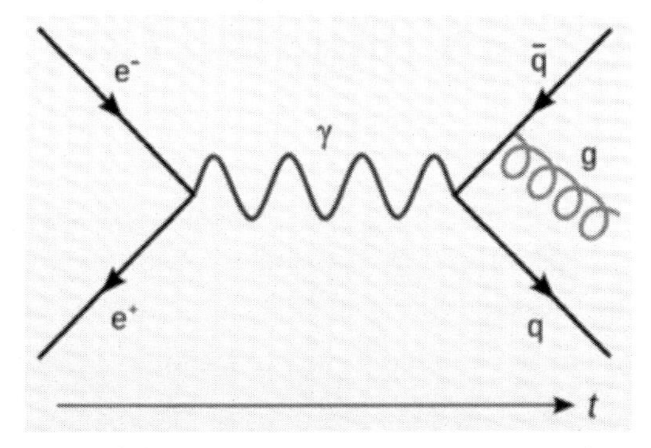

Figure 3. The Feynman diagram depicting the electron-positron annihilation and the gluon emission, represented by a helix.

In the case of a proton made up of three independent quarks, each of them contributes a third of the proton's momentum. If the three quarks communicate via the exchange of gluons, they transfer momentum to each other. It means that the quarks can have higher or lower momentum fractions and their structure function broadens. The gluons themselves are responsible for about half of the momentum. The more quark-antiquark pairs and gluons are to be found in the proton, the more the structure function increases toward lower momentum fractions. This is a very important insight revealed by results from the HERA experiments.

The measurements of the structure functions $F_2(x,Q^2)$ by DIS processes in the small x region have opened up a new era in parton density measurements inside hadrons. The structure functions reflect the momentum distribution of partons in a nucleon. It is also important to know the gluon distribution inside a hadron at small x because gluons are expected to be dominant in this region. The steep increase in $F_2(x,Q^2)$ towards small x observed at HERA also indicates a similar increase in gluon distribution towards small x in perturbative quantum chromodynamics [15]. One of the most striking discoveries at HERA is the steep rise of the proton structure function $F_2(x,Q^2)$ with decreasing Bjorken-x [16]. The behavior of the structure function at small x is driven by the gluon through the process $g \to q\bar{q}$. Thus the gluon distribution is the observable that governs the physics of high-energy processes in QCD. Gluon interaction as represented in Feynman diagram is shown in Figure 3.

2. GLUONS AT HERA AND LHC

The most important particle colliders use hadrons, viz. HERA was an *ep* collider, the Tevatron is a $p\bar{p}$ collider and the LHC at CERN is a *pp* collider. At high energies, the cross sections of hadrons are dominated by scatterings involving gluons. Gluons clearly outnumber quarks in the small momentum fraction (small x) range of the parton distribution functions as a consequence of the QCD parton splitting probabilities described by the DGLAP [17-19] and BFKL [20-23] evolution equations. The fast growth of the gluon densities for decreasing x conspicuously observed in DIS *ep* at HERA cannot, however, continue indefinitely since this would violate unitarity. For small enough x values, gluons must start to recombine in a process known as gluon saturation [14, 24].

A particularly interesting region of strong interactions is hard scattering processes in the region where the center-of-mass energy becomes large as compared to the momentum transfer. In DIS, this limit corresponds to values of the Bjorken variable $x << 1$.

The experimental investigation of small x processes became possible with the advent of high-energy colliders (colliding beam facilities). Extensive studies of DIS at small x have been performed at the HERA electron-proton (*ep*) collider at DESY (Deutsches Elektronen Synchrotron), Hamburg. Measurements of inclusive cross sections have spectacularly confirmed the rise of the gluon density in the proton at small x, as predicted by QCD, down to values $x \sim 10^{-4}$.

Another potentially much more powerful laboratory for studying small-x physics is the high-energy proton-proton (*pp*) collider: the LHC at CERN. At LHC, the hard processes can probe parton distributions down to values of $x \sim 10^{-7}$ [25]. In *pp* scattering, as compared to *ep*, one is dealing with collisions of two objects with a complex internal structure.

This results, e.g., in a high probability of multiple hard scattering processes at high energies, and a much richer spectrum of soft hadronic interactions.

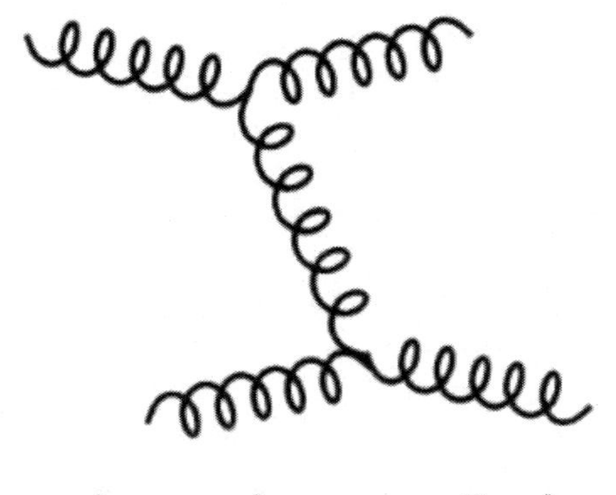

Figure 4. Gluon emission and splitting as the proton structure is probed deeper and deeper, i.e., at low Bjorken-x.

Thus, although QCD factorization can still be applied to certain hard processes in *pp* collisions, the modeling of the hadronic environment of the quarks and gluons participating in the hard processes is generally much more challenging than in *ep* scattering [26].

3. Self-Similarity Based Gluon Distribution Function

As mentioned in the earlier section, in QCD the behavior of the sea quark densities is driven by gluon emissions and splittings. The deeper the proton structure is probed, the more gluon-gluon interactions can be observed (see Figure 4). These, in analogy to fractals, may follow self-similarity, i.e., scaling described by a power law [27-30].

The study of structure functions at small x has become topical in view of the high energy collider like HERA where previously unexplored small x regime is being reached. The measurement of the longitudinal structure function $F_L(x,Q^2)$ is of great theoretical importance, since it may allow us to distinguish between different models describing the QCD evolution at small x. From experimental determinations by H1 [31-33], which used assumptions on the behavior of $F_2(x,Q^2)$ in extracting $F_L(x,Q^2)$, and from theoretical analyses of the inclusive DIS cross section data [34-36], the longitudinal structure function at small x is expected to be significantly larger than zero. It is, therefore, vital to have an accurate measurement of $F_L(x,Q^2)$ at HERA. Its experimental determination is usually difficult since it usually requires cross section measurements at different values of the center-of-mass energy implying change of beam energies. However, the H1 Collaboration at HERA [37] has extracted experimental results on $F_L(x,Q^2)$ by measuring the cross section in a kinematical region where its contribution is substantial.

Scaling violations, attributed to the high gluon density in the proton, have been observed in fixed-target DIS experiments, i.e., the variation at fixed value of Bjorken-x of the structure function with Q^2, the squared four-momentum transfer between lepton and nucleon. Scaling violations in the evolution of $F_2(x,Q^2)$ at small x, as described by the DGLAP QCD evolution equations [17-19], have previously been used to constrain the gluon distribution and $F_L(x,Q^2)$ [32].

It is important to know the gluon distribution inside a hadron at small x because gluons are expected to be dominant in this region. The steep increase in $F_2(x,Q^2)$ towards small x observed at HERA also indicates a similar increase in gluon distribution towards small x in perturbative QCD. Accurate knowledge of gluon distribution function at small x and large virtuality Q^2 plays a vital role in estimating QCD backgrounds and in calculating gluon-initiated processes, and thus in our ability to search for new physics at the LHC. The gluon and quark distribution functions have traditionally been determined simultaneously by fitting experimental data on neutral and charged current deep inelastic scattering processes and some jet data over a large domain of values of x and Q^2. The distributions at small x and large Q^2 are determined mainly by the proton structure function $F_2(x,Q^2)$ measured in deep inelastic *ep* scattering [38]. The exact relation between the gluon distribution and the quark distribution is not derivable in QCD even in leading order (LO).

CONCLUSION

The quark-parton model had to be extended, and became the field theory of QCD, in which the gluons are field carriers, just like photons in QED. There are eight kinds of gluons that are characterized in terms of a new quantum number called color, which is carried by both quarks and the gluons themselves, in contrast to QED, where the field carrier is uncharged. The gluon can thus interact with itself as well as with quarks.

As with all quantum phenomena, what is in a proton depends upon how one looks at it. A more energetic probe has a smaller wavelength and, therefore, can reveal smaller structures, but it also injects energy into the system, and this allows the creation of new particles. At higher scales, quarks radiate gluons that then split into quark-antiquark pairs, which again radiate gluons: and the gluons themselves can also radiate gluons. The valence quarks thus lose momentum, distributing it between the sea quarks and gluons - increasingly many, with smaller and smaller amounts of momentum [39].

In fact, thc HERA data already give hints that we may be entering a new phase of QCD at very small x, where the gluon density is very large. Such large densities could lead to nonlinear effects in which gluons recombine. When the rate of recombination equals the rate of gluon splitting, one may get gluon saturation. This state of matter has been described as a color glass condensate (CGC) and has been further probed in heavy-ion experiments at the LHC and at RHIC at Brookhaven National Laboratory. The higher gluon densities involved in experiments with heavy nuclei enhance the impact of nonlinear gluon interactions.

REFERENCES

[1] Gell-Mann, M. (1964). A schematic model of baryons and mesons. *Physics Letters*, 8, 214-215.

[2] Zweig, G. (1964). An SU(3) model for strong interaction symmetry and its breaking. *Preprint CERN Report* No. 8419/TH412.

[3] Taylor, R. E. (1991). Deep inelastic scattering: The early years, *Reviews of Modern Physics*, 63, 573-595.

[4] Kendall, H. W. (1991). Deep inelastic scattering: Experiments on the proton and the observation of scaling, *Reviews of Modern Physics*, 63, 597-614.

[5] Friedman, J. I. (1991). Deep inelastic scattering: Comparisons with the quark model, *Reviews of Modern Physics*, 63, 615-629.

[6] Cooper-Sarkar, A. M., Devenish, R. C. E. and De Roeck, A. (1998). Structure functions of the nucleon and their interpretation, *International Journal of Modern Physics A*, 13, 3385-358.

[7] Feynman, R. P. (1969). Very high-energy collision of hadrons, *Physical Review Letters*, 23, 1415-1417.

[8] Bjorken, J. D. (1969). Asymptotic sum rules at infinite momentum, *Physical Review*, 179, 1547-1553.

[9] Bjorken, J. D. and Paschos, E. A. (1969). Inelastic electron-proton and γ-proton scattering and the structure of the nucleon, *Physical Review*, 185, 1975-1982.

[10] Bjorken, J. D. and Paschos, E. A. (1970). High-energy inelastic neutrino-nucleon interactions, *Physical Review D*, 1, 3151-3160.

[11] Gross, D. J. and Wilczek, F. (1973). Ultraviolet behavior of non-Abelian gauge theories, *Physical Review Letters*, 30, 1343-1346.

[12] Politzer, H. D. (1973). Reliable perturbative results for strong interactions?, *Physical Review Letters*, 30, 1346-1349.

[13] Adloff, C. et al. (H1 Collaboration). (1999). Charged particle cross sections in photoproduction and extraction of the gluon density in the photon, *European Physical Journal C*, 10, 363-372.

[14] Gribov, L. V., Levin, E. M. and Ryskin, M. G. (1983). Semihard processes in QCD, *Physics Reports*, 100, 1-150.

[15] Rezaei, B., Boroun, G. R. and Teimoury, F. (2012). An analysis of the proton structure function from the gluon distribution function, *Physica Scripta*, 86, 015101.

[16] Aid, S. et al. (H1 Collaboration). (1996). Strangeness production in deep inelastic positron-proton scattering at HERA, *Nuclear Physics B*, 480, 3-34.

[17] Altarelli, G. and Parisi, G. (1977). Asymptotic freedom in parton language, *Nuclear Physics B*, 126, 298-318.

[18] Gribov, V. N. and Lipatov, L. N. (1972). Deep inelastic ep scattering in perturbation theory, *Soviet Journal of Nuclear Physics*, 15, 438-450.

[19] Dokshitzer, Y. L. (1977). Calculation of the structure functions for deep inelastic scattering and e^+e^- annihilation by perturbation theory in Quantum Chromodynamics, *Soviet Physics - Journal of Experimental and Theoretical Physics*, 46, 641-653.

[20] Fadin, V. S., Kuraev, E. A. and Lipatov, L. N. (1975). On the Pomeranchuk singularity in Asymptotically free theories, *Physics Letters B*, 60, 50-52.

[21] Kuraev, E. A., Lipatov, L. N. and Fadin, V. S. (1976). Multi-Reggeon processes in the Yang-Mills theory, *Soviet Physics - Journal of Experimental and Theoretical Physics*, 44, 443-450.

[22] Kuraev, E. A., Lipatov, L. N. and Fadin, V. S. (1977). The Pomeranchuk singularity in nonabelian gauge theories, *Soviet Physics - Journal of Experimental and Theoretical Physics*, 45, 199-204.

[23] Balitsky, I. I. and Lipatov, L. N. (1978). The Pomeranchuk singularity in Quantum Chromodynamics, *Soviet Journal of Nuclear Physics*, 28, 822-829.

[24] Mueller, A. H. and Qiu, J. (1986). Gluon recombination and shadowing at small values of *x*, *Nuclear Physics B*, 268, 427-452.

[25] *Talk by Nadolsky, P. M.* (2013). Peking University, Beijing, China.

[26] Frankfurt, L., Strikman, M. and Weiss, C. (2005). Small-*x* physics: From HERA to LHC and beyond, *Annual Review of Nuclear and Particle Science*, 55, 403-465.

[27] Lastovicka, T. (2002). Self-similar properties of the proton structure at low *x*, *European Physical Journal C*, 24, 529-533.

[28] Akbari Jahan and Choudhury, D. K. (2011). Fractal inspired models of quark and gluon distributions and longitudinal structure function $F_L(x,Q^2)$ at small x, *Indian Journal of Physics*, 85, 587-596.
[29] Akbari Jahan and Choudhury, D. K. (2012). Momentum fractions of quarks and gluons in a self-similarity based model of proton, *Modern Physics Letters A*, 27, 1250193.
[30] Akbari Jahan and Choudhury, D. K. (2013). An analysis of momentum fractions of quarks and gluons in a model of proton, *Modern Physics Letters A*, 28, 1350086.
[31] Adloff, C. et al. (H1 Collaboration). (2001). Deep inelastic inclusive ep scattering at low x and a determination of α_s, *European Physical Journal C*, 21, 33-61.
[32] Adloff, C. et al. (H1 Collaboration). (1997). Determination of the longitudinal proton structure function $F_L(x,Q^2)$ at low x, *Physics Letters B*, 393, 452-464.
[33] Adloff, C. et al. (H1 Collaboration). (2003). Measurement and QCD analysis of neutral and charged current cross sections at HERA, *European Physical Journal C*, 30, 1-32.
[34] Martin, A. D., Stirling, W. J., Thorne, R. S. and Watt, G. (2007). Update of parton distributions at NNLO, *Physics Letters B*, 652, 292-299.
[35] Pumplin, J., Lai, H. L. and Tung, W. K. (2007). Charm parton content of the nucleon, *Physical Review D*, 75, 054029.
[36] Nadolsky, P. M. et al. (2008). Implications of CTEQ global analysis for collider observables, *Physical Review D*, 78, 013004.
[37] Nelly Gogitidze (2002). Determination of the longitudinal structure function F_L at HERA, *Journal of Physics G: Nuclear and Particle Physics*, 28, 751-765.
[38] Block, M. M., Durand, L., Ha, P. and McKay, D. W. (2011). An analytic solution to LO couples DGLAP evolution equations: a new pQCD tool, *Physical Review D*, 83, 054009.
[39] Devenish, R. and Cooper-Sarkar, A. (2004). *Deep Inelastic Scattering*, Oxford University Press, Oxford.

In: Horizons in World Physics. Volume 302 ISBN: 978-1-53616-472-5
Editor: Albert Reimer

Chapter 8

DIRECT IMPLICIT SCHEMES FOR PROBLEMS OF LINEAR ADVECTION-DIFFUSION AND NONLINEAR DIFFUSION ON A SPHERE

Yuri N. Skiba[1,*] ***Denis M. Filatov***[2,†]
and Roberto C. Cruz-Rodríguez[1,‡]
[1]Centro de Ciencas de la Atmósfera,
Universidad Nacional Autonóma de México,
Mexico City, Mexico
[2]Sceptica Scientific Ltd, Stockport, Cheshire, UK

Abstract

Three methods for solving the problems of linear advection-diffusion and nonlinear diffusion on a sphere are proposed.

The first method is developed for both of these problems. The velocity field on the sphere is assumed to be non-divergent and known. Discretisation of the advection-diffusion equation in space is performed by the finite volume method using the Gauss theorem for every grid cell. For the discretisation in time the symmetrised dicyclic component-wise

*Corresponding Author's E-mail: skiba@unam.mx
†Corresponding Author's E-mail: denis.filatov@sceptica.co.uk
‡Corresponding Author's E-mail: roberto.cruz.rdg@gmail.com

splitting method and the Crank-Nicolson schemes are used. The one-dimensional periodic problems arising at splitting in the longitudinal direction are solved via Sherman-Morrison's formula and by Thomas' algorithm. The highlight of the method is the use of special bordered matrices for direct (i.e., non-iterative) solving the 1D problems arising at splitting in the latitudinal direction. The bordering procedure requires a prior determination of the solution at the poles. The resulting linear systems have tridiagonal matrices and are solved by Thomas' algorithm, which ensures the second approximation order in space and time. The method is thus implicit, unconditionally stable, non-iterative and computationally cheap. The theoretical results are confirmed numerically by simulating various linear advection-diffusion problems and nonlinear diffusion processes. The numerical tests show high accuracy and efficiency of the method that correctly describes the advection-diffusion processes and the mass balance of a substance in a forced and dissipative discrete system. In addition, in the absence of external forcing and dissipation, it conserves both the total mass and the norm (or energy) of solution.

To solve nonlinear diffusion problems on a sphere, apart from the pole-bordering method two implicit, balanced and unconditionally stable finite-difference schemes of the second and fourth approximation orders in spatial variables are proposed. Unlike the first, pole-bordering method, both nonlinear solvers exclude from consideration the polar grid cells with the poles, thereby avoiding the challenge of imposing suitable boundary conditions. The highlight of these two methods is the use of different coordinate maps of the sphere at each stage of splitting. As a result, all one-dimensional split problems are solved with periodic boundary conditions in both directions — in latitude and in longitude. In particular, all the 1D split problems in the first nonlinear method are solved using Sherman-Morrison's formula and Thomas' algorithm.

The common part of all the three methods is operator splitting. Due to the dicyclic coordinate splitting all the methods possess the second approximation order in time. The operator splitting also ensures direct and computationally cheap implementation of all the implicit schemes, as well as allows using parallel processors when solving the corresponding 1D split equations.

Keywords: advection-diffusion problem, nonlinear diffusion problem, finite volume method, splitting method, bordering method, direct implicit algorithm.

1. INTRODUCTION

Currently there is extensive literature on numerical methods for solving the advection-diffusion and nonlinear diffusion problems (see, for example, [1–12]). However, in many papers boundary value problems are mostly studied, while much less research is dedicated to the numerical simulation of advection-diffusion and nonlinear diffusion processes on closed manifolds that have no boundaries (e.g., sphere). The work is aimed at filling this gap up.

The first part of this paper describes and tests a balanced, implicit, unconditionally stable second-order numerical algorithm proposed in [13]. It demonstrates the algorithms ability to describe linear advection-diffusion and nonlinear diffusion processes in the spherical geometry. Throughout the paper we call it *Method 1*. When considering linear advection-diffusion problems, the velocity field on the sphere is non-divergent and assumed to be known. Discretisation of the advection-diffusion equation in space is performed by the finite volume method using the Gauss theorem for each grid cell. One of the cores of Method 1 is operator splitting. For the discretisation in time, the symmetrised dicyclic component-wise splitting method and the Crank-Nicolson schemes are used [13–15]. The one-dimensional periodic problems arising at splitting in the longitudinal direction are solved with Sherman-Morrison's formula [16,17] and Thomas' algorithm [18]. Also, special bordered matrices are used for direct (non-iterative) solving the 1D problems arising at splitting in the latitudinal direction. The bordering method requires a prior determination of the solution at the poles. The resulting linear systems in each meridional interval between the poles have tridiagonal matrices and are solved by Thomas' algorithm. Due to the use of the dicyclic coordinate splitting the method has second approximation order in time. The operator splitting also allows non-iterative and computationally efficient implementation of the implicit schemes, as well as permits using parallel processors when solving the 1D split equations. The algorithm is of second approximation order in space.

The results of various tests show that the developed numerical model correctly simulates the processes of advection and diffusion on the entire sphere including the areas around the poles. Therefore, it is useful for solving a variety of meteorological and physical problems, in particular the transport of quasi-passive pollutants, temperature and water vapor in the Earths atmosphere, as well as solving combustion problems associated with the propagation of nonlinear temperature waves, simulating various blow-up regimes, among others. Yet, it can efficiently solve the linearised barotropic vorticity equation [19], as well as some elliptic and adjoint advection-diffusion problems [13]. Finally, the method can be the spherical part of a more general splitting algorithm aimed at solving three-dimensional advection-diffusion problems in a spherical shell, which often arise in meteorology, geophysics, plasma physics, chemical kinetics and ecology [10].

The second part of the research is devoted to describing two methods for solving various problems of mostly nonlinear diffusion on a sphere.

The phenomenon of diffusion has many manifestations in nature. One of the most obvious is from gas dynamics and atmospheric environment. Air, as it is known, is a mixture of different gases, such as carbon dioxide, oxygen, hydrogen, nitrogen and particles of dust. However, due to diffusion the atmospheric composition at any given altitude is fairly homogeneous. Diffusion plays an important role in plant nutrition, transport of nutrients and oxygen in humans and animals as well. In the food industry it is widely used for preserving fruits and vegetables, pickling cucumbers. Diffusion is used in the electronics industry. Many semiconductor devices have been made with the help of diffusion. Diffusion is invoked in social sciences to describe the spread of ideas. Overall, diffusion is relevant to a wide range of important physical phenomena and has many practical applications.

Despite the atomic-molecular theory, it is often convenient to abstract from the discreteness of space and time while describing the diffusion phenomenon. The diffusion process can be regarded as the fluid motion in a continuous medium. Hence, there are numerous analogies which can help clarify a number of features of diffusion, and first of all, the analogy between diffusion and thermal conductivity. Since these processes are described by the same differential equations, one can use ready-made solutions from the more developed theory of heat conduction by replacing the coefficient of thermal diffusivity with the diffusion coefficient, and the temperature with the concentration. This approach is widely used in practice. The main objective of the analytic theory of diffusion is to study the spatial and temporal changes in the basic physical quantity that characterises the process of diffusion, be it the concentration of a substance, the temperature of a medium, etc. The laws, governing the spatial and temporal development of the concentration field, are called Ficks laws [20].

A large number of important natural phenomena, e.g., heat transfer in ionised gases, gas percolation through porous media, concentration waves in distributed chemical reactors, combustion, viscous processes in diverse media, as well as many others are described by *nonlinear* diffusion equations [14, 21–29]. For example, the nonlinear diffusion equations, widely used in the mathematical theory of combustion, deal with a combination of equations of chemical kinetics on the one hand, and the thermal conductivity and diffusion — on the other. The reaction rate always depends on the temperature in an essentially nonlinear way. This nonlinearity is an important feature of the phenomena of combustion; without the nonlinearity critical conditions of the combustion disappear, and the concept of combustion loses its meaning [30]. The whole theory of combustion is based on the assumption that the burning rate depends rather on the temperature than on all other parameters. Thermal effect of reaction must also be large.

In some practical applications the diffusion problem has to be studied on a sphere. For example, the diffusion equation on a spherical surface was considered in [31] for the quantitative study of the translational diffusion in the membranes of single cells (see also [32]). Another classical problem is the heat or pollution transfer in the atmospherc. A more sophisticated example is the single-particle Schrdinger equation which, formally speaking, is diffusion-like. The increased interest in recent years has been observed to the spiral wave solutions to reaction-diffusion equations on a sphere as well [33, 34].

The development of numerical methods for solving the diffusion equation on the sphere has specific features. The point is that the sphere is not a doubly periodic domain, since it is periodic in the longitude, but is not in the latitude due to the presence of two poles. Therefore, a numerical procedure designed to solve the original 2D problem will be computationally cumbersome, because the matrix of the resulting linear system will be of a general type, thereby not permitting to apply fast linear solvers. As for making some or other modifications to the model prior to computing, these normally bring to the necessity of constructing mathematically and physically correct boundary conditions or involving special numerical procedures near the poles (e.g., matrix bordering, as mentioned above with regard to Method 1). Both are always a challenge, because the poles represent an artificial boundary appearing exclusively due to the latitudinal-longitudinal coordinate system, while the construction of proper artificial boundary conditions is a serious independent question [35].

Some of the three-dimensional nonlinear diffusion equations considered in the spherical geometry are easier to solve by separating the diffusion operator into the spherical and radial components. Since the spherical nonlinear diffusion component is more complicated, the second part of this research is devoted to two advanced numerical methods for the solution of nonlinear diffusion equations on a sphere (called *Methods 2* and *3*).

Unlike Method 1, Methods 2 and 3 exclude from consideration both polar grid cells with the poles. The highlight of these methods is the use of two different coordinate maps of the sphere at each stage of the splitting. The key advantage of this technique is that, although the sphere is not a doubly periodic domain, each split 1D equation can be equipped with a periodic boundary condition in the corresponding (either latitudinal or longitudinal) direction. Therefore, unlike other existing methods, these ones do not require applying special numerical procedures for careful computing the solution near the poles, which is always a challenge. For example, Method 2 possesses the second approximation order in space, and thus all its problems are solved using Sherman-Morrison's formula and Thomas' algorithm.

The common point of Methods 1 to 3 is operator splitting. Due to the dicyclic coordinate splitting all the three methods have the second approximation order in time. Operator splitting also ensures direct and cost-efficient computing of all the implicit schemes, as well as permits using parallel processors when solving the 1D split equations. Last but not least, it is to emphasise that although Method 1 qualitatively differs from Methods 2 and 3, both approaches allow accurate covering the whole sphere, and hence provide solution in the entire computational domain.

2. LINEAR ADVECTION-DIFFUSION PROBLEM

Let $\phi(\mathbf{x}, t)$ be the concentration of a physical substance at the point $\mathbf{x} = (\lambda, \vartheta)$ of the sphere S of radius a, and let $\mathbf{U} = \{u(\mathbf{x}, t), v(\mathbf{x}, t)\}$ be a known non-divergent velocity field on S. The advection-diffusion problem is formulated as follows

$$\phi_t + A\phi + L\phi + \sigma\phi = f, \qquad \phi(\mathbf{x}, 0) = \phi^0(\mathbf{x}), \tag{1}$$

where

$$A\phi = \mathrm{div}(\mathbf{U}\phi), \qquad L\phi = -\mathrm{div}(\mu\nabla\phi), \tag{2}$$

while

$$\mathrm{div}\mathbf{U} = \frac{1}{a\sin\vartheta}[u_\lambda + (v\sin\vartheta)_\vartheta] = 0. \tag{3}$$

Here λ is the longitude and ϑ is the colatitude (the difference between 90° and the latitude $\varphi = 90^\circ - \vartheta$), $\mu(\mathbf{x}, t) > 0$ is the diffusion coefficient, ∇ is the gradient along the surface of the sphere, $\sigma(\mathbf{x}, t) > 0$ characterises the rate of exponential decay of $\phi(\mathbf{x}, t)$ due to physical and chemical processes, and $f(\mathbf{x}, t)$ is a known forcing (for example, the intensity of pollution sources). Hereinafter $(\cdot)_t$, $(\cdot)_\lambda$ and $(\cdot)_\vartheta$ denote partial derivatives in t, λ and ϑ, respectively, unless otherwise specified.

Note that A is a skew-symmetric operator and L is a positive semidefinite operator. Indeed, the inner product of two functions on the sphere is defined as $\langle\phi, g\rangle = \int_S \phi g dS$, and since $\int_S \nabla\cdot\mathbf{u}dS = 0$ for any vector-function $\mathbf{u}$ on the sphere, integration of (2) by parts, given (3), leads to

$$\begin{aligned}
\langle A\phi, \phi\rangle &= \int_S \phi\nabla\cdot(\mathbf{U}\phi)dS = -\frac{1}{2}\int_S \nabla\cdot(\mathbf{U}\phi^2)dS = 0,\\
\langle L\phi, \phi\rangle &= -\int_S \phi\nabla\cdot(\mu\nabla\phi)dS = \int_S \mu|\nabla\phi|^2 dS \geq 0.
\end{aligned}$$

The total mass $\int_S \phi dS$ of substance satisfies the balance equation

$$\frac{\partial}{\partial t}\int_S \phi dS = \int_S f dS - \int_S \sigma\phi dS, \tag{4}$$

while the evolution of the L_2-norm $\|\phi\| = \langle\phi, \phi\rangle^{\frac{1}{2}}$ of ϕ is governed by the integral equation

$$\frac{1}{2}\frac{\partial}{\partial t}\int_S \phi^2 dS = \int_S f\phi dS - \int_S (\sigma\phi^2 + \mu|\nabla\phi|^2)dS, \tag{5}$$

where $dS = a^2 \sin\vartheta d\lambda d\vartheta$. Thus, the total mass of the substance and the norm of solution $\|\phi\|$ (or solution's energy) grow under the influence of the external forcing ($f \neq 0$) and decrease because of the decay process ($\sigma \neq 0$). Besides, the norm $\|\phi\|$ additionally decreases due to diffusion ($\mu \neq 0$). In particular, if $f = \mu = \sigma = 0$ then both characteristics are conserved in time —

$$\frac{\partial}{\partial t}\int_S \phi dS = 0, \qquad \frac{\partial}{\partial t}\|\phi\| = 0. \tag{6}$$

3. Finite Volume Method (Method 1)

3.1. Discretisation

We use a spherical grid with constant steps $\Delta\lambda$ and $\Delta\vartheta$ in the longitudinal and colatitudinal directions, respectively. The sphere's surface is partitioned into $I \times J+2$ non-overlapping boxes, or grid cells, as follows: $S = \bigcup_{ij} S_{ij} \bigcup S_0 \bigcup S_{J+1}$, where S_0 and S_{J+1} are the round cells centered at the North ($0 \leq \vartheta \leq \vartheta_{1/2} = \Delta\vartheta/2$) and South pole ($\pi - \Delta\vartheta/2 = \vartheta_{J+1/2} \leq \vartheta \leq \vartheta_{J+1} = \pi$), respectively, while S_{ij} is the non-pole cell, with $\lambda_i - \Delta\lambda/2 \leq \lambda \leq \lambda_i + \Delta\lambda/2$ and $\vartheta_j - \Delta\vartheta/2 \leq \vartheta \leq \vartheta_j + \Delta\vartheta/2$, centered at the point (λ_i, ϑ_j) with the area $|S_{ij}| = a^2\Delta\lambda\Delta\vartheta \sin\vartheta_j$, $i = \overline{1, I}$, $j = \overline{1, J}$ (see Figure 1). The radius of the round cells S_0 and S_{J+1} is $a\Delta\vartheta/2$, and hence the spherical area of each pole cell is $|S_0| = |S_{J+1}| = \pi a^2(\Delta\vartheta)^2/4$.

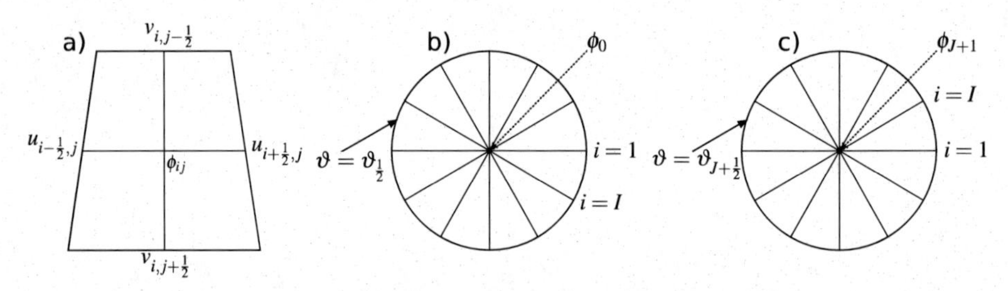

Figure 1. Partition of the sphere by grid cells: (a) a non-pole grid cell; (b) the North-pole grid cell; (c) the South-pole grid cell.

The discretisation of each term of equation (1) at the node of a grid cell S_Δ is performed using the finite volume method, i.e., via the Gauss theorem applied to the mean value of this term in the cell —

$$\int_{S_\Delta} \nabla \cdot \mathbf{P} dS = \oint_\Gamma \mathbf{P} \cdot \mathbf{n} dl,$$

where $\mathbf{P}$ is the corresponding vector field, Γ is the contour bounding the cell S_Δ, where Δ stands for either ij or 0 or $J+1$. It is easy to show that all the approximations are of the second-order accuracy with respect to $\Delta\lambda$ and $\Delta\vartheta$.

The discrete form of the continuity equation (3) is

$$\frac{1}{a \sin \vartheta_j} \left(\frac{u_{i+\frac{1}{2},j} - u_{i-\frac{1}{2},j}}{\Delta\lambda} + \frac{v_{i,j+\frac{1}{2}} \sin \vartheta_{j+\frac{1}{2}} - v_{i,j-\frac{1}{2}} \sin \vartheta_{j-\frac{1}{2}}}{\Delta\vartheta} \right) = 0 \quad (7)$$

at any non-pole node (λ_i, ϑ_j), and

$$\sum_{i=1}^{I} v_{i,1/2} = 0, \qquad \sum_{i=1}^{I} v_{i,J+1/2} = 0 \quad (8)$$

at the North and South poles, respectively. To discretise the advection term, we use the discrete continuity equations (7)-(8) and the second-order approximation formulas

$$\phi_{i\pm1/2,j} = \frac{\phi_{i\pm1,j} + \phi_{i,j}}{2}, \qquad \phi_{i,j\pm1/2} = \frac{\phi_{i,j\pm1} + \phi_{i,j}}{2},$$
$$\phi_{i,1/2} = \frac{\phi_0 + \phi_{i,1}}{2}, \qquad \phi_{i,J+1/2} = \frac{\phi_{J+1} + \phi_{i,J}}{2}. \quad (9)$$

As a result, we obtain

$$\nabla \cdot (\mathbf{U}\phi)|_{ij} \approx \frac{1}{|S_{ij}|} \int_{S_{ij}} \nabla \cdot (\mathbf{U}\phi) dS \approx \frac{\phi_{i+1,j} u_{i+\frac{1}{2},j} - \phi_{i-1,j} u_{i-\frac{1}{2},j}}{2a\Delta\lambda \sin \vartheta_j}$$
$$+ \frac{\phi_{i,j+1} v_{i,j+\frac{1}{2}} \sin \vartheta_{j+\frac{1}{2}} - \phi_{i,j-1} v_{i,j-\frac{1}{2}} \sin \vartheta_{j-\frac{1}{2}}}{2a\Delta\vartheta \sin \vartheta_j}, \quad (10)$$

$$\nabla \cdot (\mathbf{U}\phi)|_{\vartheta=0} \approx \frac{1}{|S_0|} \int_{S_0} \nabla \cdot (\mathbf{U}\phi) dS \approx \frac{2}{Ia\Delta\vartheta} \sum_{i=1}^{I} v_{i,\frac{1}{2}} \phi_{i,1}, \quad (11)$$

$$\nabla\cdot(\mathbf{U}\phi)|_{\vartheta=\pi}\approx\frac{1}{|S_{J+1}|}\int_{S_{J+1}}\nabla\cdot(\mathbf{U}\phi)dS\approx-\frac{2}{Ia\Delta\vartheta}\sum_{i=1}^{I}v_{i,J+\frac{1}{2}}\phi_{i,J}. \quad (12)$$

The second-order approximations of the turbulent diffusion term are written as

$$\begin{aligned}\nabla\cdot(\mu\nabla\phi)|_{ij} &\approx \frac{1}{|S_{ij}|}\int_{S_{ij}}\nabla\cdot(\mu\nabla\phi)dS\\ &\approx \frac{\mu_{i+\frac{1}{2},j}(\phi_{i+1,j}-\phi_{ij})-\mu_{i-\frac{1}{2},j}(\phi_{ij}-\phi_{i-1,j})}{(\Delta\lambda)^2a^2\sin^2\vartheta_j}\\ &+ \frac{\mu_{i,j+\frac{1}{2}}(\phi_{i,j+1}-\phi_{ij})\sin\vartheta_{j+\frac{1}{2}}-\mu_{i,j-\frac{1}{2}}(\phi_{ij}-\phi_{i,j-1})\sin\vartheta_{j-\frac{1}{2}}}{(\Delta\vartheta)^2a^2\sin\vartheta_j}\end{aligned} \quad (13)$$

$$\nabla\cdot(\mu\nabla\phi)|_{\vartheta=0}\approx\frac{1}{|S_0|}\int_{S_0}\nabla\cdot(\mu\nabla\phi)dS\approx\frac{4}{Ia^2(\Delta\vartheta)^2}\sum_{i=1}^{I}\mu_{i,\frac{1}{2}}(\phi_{i,1}-\phi_0), \quad (14)$$

$$\begin{aligned}\nabla\cdot(\mu\nabla\phi)|_{\vartheta=\pi} &\approx \frac{1}{|S_{J+1}|}\int_{S_{J+1}}\nabla\cdot(\mu\nabla\phi)dS\\ &\approx -\frac{4}{Ia^2(\Delta\vartheta)^2}\sum_{i=1}^{I}\mu_{i,J+\frac{1}{2}}(\phi_{J+1}-\phi_{i,J}).\end{aligned} \quad (15)$$

3.2. Discrete Model

Let $\vec{\phi}=\{\phi_0,\phi_{ij},\phi_{J+1}\}^T$ be the vector of numerical solution and let $\vec{f}=\{f_0,f_{ij},f_{J+1}\}^T$ be the vector formed by values of the forcing f, both defined in the grid nodes, where $i=\overline{1,I}$, $j=\overline{1,J}$. Thus, the dimension of each vector is $I\times J+2$. Then the discrete advection-diffusion problem can be written in the matrix form as

$$\frac{d}{dt}\vec{\phi}+\sum_{k=1}^{2}R_k\vec{\phi}+\sigma\vec{\phi}=\vec{f}, \quad (16)$$

where $R_k=A_k+L_k$ are the matrices obtained during the discretisation of the advection operator A (the matrices A_k) and the turbulent diffusion operator L (the matrices L_k) of equation (1) in the λ- ($k=1$) and ϑ-direction ($k=2$), respectively, —

$$R_1\vec{\phi} = \begin{cases} 0\,, & \text{if } \vartheta = 0 \text{ (North pole)} \\ (R_1\vec{\phi})_{ij}\,, & i = \overline{1,I}, j = \overline{1,J} \\ 0\,, & \text{if } \vartheta = \pi \text{ (South pole)} \end{cases}, \tag{17}$$

$$R_2\vec{\phi} = \begin{cases} (R_2\vec{\phi})_0\,, & \text{if } \vartheta = 0 \text{ (North pole)} \\ (R_2\vec{\phi})_{ij}\,, & i = \overline{1,I}, j = \overline{1,J} \\ (R_2\vec{\phi})_{J+1}\,, & \text{if } \vartheta = \pi \text{ (South pole)} \end{cases}, \tag{18}$$

$$(A_1\vec{\phi})_{ij} = \frac{1}{a\sin\vartheta_j}\frac{u_{i+\frac{1}{2},j}\phi_{i+1,j} - u_{i-\frac{1}{2},j}\phi_{i-1,j}}{2\Delta\lambda}, \tag{19}$$

$$(L_1\vec{\phi})_{ij} = -\frac{1}{a^2\sin^2\vartheta_j}\frac{\mu_{i+\frac{1}{2},j}(\phi_{i+1,j} - \phi_{ij}) - \mu_{i-\frac{1}{2},j}(\phi_{ij} - \phi_{i-1,j})}{(\Delta\lambda)^2}, \tag{20}$$

$$(A_2\vec{\phi})_0 = \frac{2}{Ia\Delta\vartheta}\sum_{i=1}^{I} v_{i,\frac{1}{2}}\phi_{i,1}, \qquad (L_2\vec{\phi})_0 = \frac{-4}{Ia^2(\Delta\vartheta)^2}\sum_{i=1}^{I}\mu_{i,\frac{1}{2}}(\phi_{i,1} - \phi_0), \tag{21}$$

$$(A_2\vec{\phi})_{ij} = \frac{1}{a\sin\vartheta_j}\frac{v_{i,j+\frac{1}{2}}\phi_{i,j+1}\sin\vartheta_{j+\frac{1}{2}} - v_{i,j-\frac{1}{2}}\phi_{i,j-1}\sin\vartheta_{j-\frac{1}{2}}}{2\Delta\vartheta}, \tag{22}$$

$$(L_2\vec{\phi})_{ij} = \frac{\mu_{i,j-\frac{1}{2}}(\phi_{ij} - \phi_{i,j-1})\sin\vartheta_{j-\frac{1}{2}} - \mu_{i,j+\frac{1}{2}}(\phi_{i,j+1} - \phi_{ij})\sin\vartheta_{j+\frac{1}{2}}}{(\Delta\vartheta)^2 a^2\sin\vartheta_j}, \tag{23}$$

$$(A_2\vec{\phi})_{J+1} = -2\sum_{i=1}^{I}\frac{v_{i,J+\frac{1}{2}}\phi_{i,J}}{Ia\Delta\vartheta}, \qquad (L_2\vec{\phi})_{J+1} = 4\sum_{i=1}^{I}\frac{\mu_{i,J+\frac{1}{2}}(\phi_{J+1} - \phi_{i,J})}{Ia^2(\Delta\vartheta)^2}. \tag{24}$$

Let us introduce the scalar product of two vectors $\vec{\phi}$ and $\vec{f}$ as follows

$$\langle\vec{\phi}, \vec{f}\rangle = \phi_0 f_0|S_0| + \sum_{i=1}^{I}\sum_{j=1}^{J}\phi_{ij}f_{ij}|S_{ij}| + \phi_{J+1}f_{J+1}|S_{J+1}|, \tag{25}$$

where $|S_{ij}| = a^2\Delta\lambda\Delta\vartheta\sin\vartheta_j$ and $|S_0| = |S_{J+1}| = \pi a^2(\Delta\vartheta)^2/4$ (see section 3.1.). The following three assertions were proved in [13]:

Lemma 1 *Let $\vec{1}$ denote the vector whose $I \times J + 2$ components are all equal to one. Then*

$$\sum_{k=1}^{2} \langle R_k \vec{\phi}, \vec{1} \rangle = 0.$$

Lemma 2 *The matrices A_k, $k = 1, 2$, are skew-symmetric.*

Lemma 3 *The matrices L_k, $k = 1, 2$, are positive semidefinite.*

Taking the scalar product of equation (16) subsequently with vectors $\vec{1}$ and $\vec{\phi}$, given Lemmas 1 and 2, we obtain the semi-discrete forms of equations (4) and (5)

$$\frac{d}{dt}\langle \vec{\phi}, \vec{1} \rangle = \langle \vec{f}, \vec{1} \rangle - \sigma \langle \vec{\phi}, \vec{1} \rangle, \tag{26}$$

$$\frac{1}{2}\frac{d}{dt}\|\vec{\phi}\|^2 = \langle \vec{f}, \vec{\phi} \rangle - \sigma \|\vec{\phi}\|^2 - \sum_{k=1}^{2} \langle L_k \vec{\phi}, \vec{\phi} \rangle, \tag{27}$$

where $\|\vec{\phi}\| = \langle \vec{\phi}, \vec{\phi} \rangle^{1/2}$.

3.3. Splitting Algorithm

In time interval $(0, T)$ let us introduce a regular grid with step τ and grid nodes $t_n = n\tau$, $n = \overline{0, 2N}$, $T = 2N\tau$. Let us also denote $\Lambda_k^n = R_k(t_n) = A_k(t_n) + L_k(t_n)$, $k = 1, 2$. In each double time interval (t_{n-1}, t_{n+1}), $n = 1, 3, 5, \ldots, 2N - 1$, equation (16) is solved by using the Marchuk symmetric dicyclic component-wise splitting algorithm [40,41]

$$\begin{aligned}
&\frac{d}{dt}\vec{\phi}_1 + \Lambda_1^n\vec{\phi}_1 = 0, \qquad \vec{\phi}_1(t_{n-1}) = \vec{\phi}(t_{n-1}) \qquad (t_{n-1} < t < t_n),\\
&\frac{d}{dt}\vec{\phi}_2 + \Lambda_2^n\vec{\phi}_2 = 0, \qquad \vec{\phi}_2(t_{n-1}) = \vec{\phi}_1(t_n) \qquad (t_{n-1} < t < t_n),\\
&\frac{d}{dt}\vec{\phi}_3 + \sigma\vec{\phi}_3 = \vec{f}, \qquad \vec{\phi}_3(t_{n-1}) = \vec{\phi}_2(t_n) \qquad (t_{n-1} < t < t_{n+1}),\\
&\frac{d}{dt}\vec{\phi}_4 + \Lambda_2^n\vec{\phi}_4 = 0, \qquad \vec{\phi}_4(t_n) = \vec{\phi}_3(t_{n+1}) \qquad (t_n < t < t_{n+1}),\\
&\frac{d}{dt}\vec{\phi}_5 + \Lambda_1^n\vec{\phi}_5 = 0, \qquad \vec{\phi}_5(t_n) = \vec{\phi}_4(t_{n+1}) \qquad (t_n < t < t_{n+1}),\\
&\qquad\qquad\qquad\qquad\qquad\qquad\qquad \vec{\phi}(t_{n+1}) = \vec{\phi}_5(t_{n+1}).
\end{aligned} \tag{28}$$

Thus, for each equation of system (28) the solution to the previous equation is used as the initial condition for the next one. The solution $\vec{\phi}_5(t_{n+1})$ approximates the solution $\vec{\phi}(t_{n+1})$ to the unsplit problem (16) and serves as the initial condition in the next time interval (t_{n+1}, t_{n+3}), etc.

Using the Crank-Nicolson scheme for the approximation of each equation (28) in time, we obtain the implicit numerical scheme

$$\begin{aligned}
\vec{\phi}\left[n-\frac{2}{3}\right] - \vec{\phi}\,[n-1] &= -\frac{\tau}{2}\Lambda_1^n\left(\vec{\phi}\left[n-\frac{2}{3}\right] + \vec{\phi}\,[n-1]\right),\\
\vec{\phi}\left[n-\frac{1}{3}\right] - \vec{\phi}\left[n-\frac{2}{3}\right] &= -\frac{\tau}{2}\Lambda_2^n\left(\vec{\phi}\left[n-\frac{1}{3}\right] + \vec{\phi}\left[n-\frac{2}{3}\right]\right),\\
\vec{\phi}\left[n+\frac{1}{3}\right] - \vec{\phi}\left[n-\frac{1}{3}\right] &= 2\tau\vec{f}[n] - \tau\sigma\left(\vec{\phi}\left[n+\frac{1}{3}\right] + \vec{\phi}\left[n-\frac{1}{3}\right]\right)\\
\vec{\phi}\left[n+\frac{2}{3}\right] - \vec{\phi}\left[n+\frac{1}{3}\right] &= -\frac{\tau}{2}\Lambda_2^n\left(\vec{\phi}\left[n+\frac{2}{3}\right] + \vec{\phi}\left[n+\frac{1}{3}\right]\right),\\
\vec{\phi}\,[n+1] - \vec{\phi}\left[n+\frac{2}{3}\right] &= -\frac{\tau}{2}\Lambda_1^n\left(\vec{\phi}\,[n+1] + \vec{\phi}\left[n+\frac{2}{3}\right]\right),
\end{aligned} \tag{29}$$

where $n = 1, 3, 5, \ldots, 2N-1$. The vectors $\vec{\phi}[n+1]$ and $\vec{\phi}[n-1]$ approximate the numerical solution $\vec{\phi}$ to problem (16) at the corresponding time moments, while $\vec{\phi}[n\pm\frac{i}{3}]$ with $i = 0, 1, 2$ are the auxiliary vectors of the splitting algorithm; $\vec{f}[n]$ is the forcing's value. The symmetrised dicyclic splitting (29) increases the approximation order of the whole model in time to $\mathcal{O}(\tau^2)$ if $\frac{\tau}{2}||\Lambda_i^n|| < 1$ for $i = 1, 2$ (see [40, sec. 5.3.3]). Using Lemmas 2 and 3, it is easy to prove that

Lemma 4 *In each finite time interval* $(0, T)$, $T = 2N\tau$, *the implicit algorithm (29) is unconditionally stable —*

$$\left\|\vec{\phi}[2n]\right\| \leq \left\|\vec{\phi}[0]\right\| + T \max_{1\leq k\leq N} \left\|\vec{f}[2k-1]\right\|, \qquad n = \overline{1, N}. \tag{30}$$

Due to Lemma 1 and given the formula $\langle \phi - \varphi, \phi + \varphi \rangle = \|\phi\|^2 - \|\varphi\|^2$, the fully discrete forms of equations (4) and (5) are

$$\langle \vec{\phi}[n+1], \vec{1} \rangle - \langle \vec{\phi}[n-1], \vec{1} \rangle = 2\tau \langle \vec{f}[n], \vec{1} \rangle - 2\tau\sigma \left\langle \frac{\vec{\phi}[n+1] + \vec{\phi}[n-1]}{2}, \vec{1} \right\rangle \tag{31}$$

and

$$\|\vec{\phi}[n+1]\|^2 - \|\vec{\phi}[n-1]\|^2 = 2\tau \langle \vec{f}[n], \vec{\phi}[n+1/3] + \vec{\phi}[n-1/3] \rangle$$
$$-\tau\sigma\|\vec{\phi}[n+1/3] + \vec{\phi}[n-1/3]\|^2 - \frac{\tau}{2}\sum_{k=1}^{2} \{\langle L_k(t_n)\vec{w}_k, \vec{w}_k \rangle + \langle L_k(t_n)\vec{z}_k, \vec{z}_k \rangle\}, \tag{32}$$

where

$$\vec{w}_1 \equiv \vec{\phi}[n-2/3] + \vec{\phi}[n-1], \qquad \vec{w}_2 \equiv \vec{\phi}[n-1/3] + \vec{\phi}[n-2/3],$$

$$\vec{z}_1 \equiv \vec{\phi}[n+1] + \vec{\phi}[n+2/3], \qquad \vec{z}_2 \equiv \vec{\phi}[n+2/3] + \vec{\phi}[n+1/3].$$

The next assertions follow from (31), (32) and Lemma 3.

Corollary 1 *If the external forcing* $\vec{f}[n]$ *is zero then the norm of solution does not grow —*

$$\left\|\vec{\phi}[n+1]\right\| \leq \left\|\vec{\phi}[n-1]\right\|. \tag{33}$$

Besides, if $\sigma = 0$ *then scheme (29), similarly to the continuous problem (1)-(3), conserves the total mass (see (6))*

$$\langle \vec{\phi}[n+1], \vec{1} \rangle = \langle \vec{\phi}[n-1], \vec{1} \rangle \tag{34}$$

Corollary 2 *If $\vec{f}[n] = 0$, $\sigma = 0$ and $\mu = 0$, then scheme (29), similarly to continuous problem (1)-(3), conserves the norm of solution, or its energy (see (6))* —

$$\left\|\vec{\phi}[n+1]\right\| = \left\|\vec{\phi}[n-1]\right\|. \tag{35}$$

Thus, scheme (29) is stable and of second approximation order in $\Delta\lambda$, $\Delta\vartheta$ and τ. Consequently, according to the Lax equivalence theorem [40,42], the following assertion holds:

Theorem 1 *If the problem is linear then numerical scheme (29) is convergent of order two with respect to $\Delta\lambda$, $\Delta\vartheta$ and τ.*

The split one-dimensional equations of system (29) are resolved by direct methods described in the following two sections.

3.3.1. Solution of the Split Systems in the λ-Direction

At the first and last stages of the splitting algorithm (29), along each fixed colatitudinal circle ϑ_j, $j = \overline{1, J}$, we have to solve a system of three-point equations of the form (the index j is omitted for simplicity)

$$a_i\phi_{i-1} - b_i\phi_i + c_i\phi_{i+1} = F_i, \qquad i = \overline{1, I}, \tag{36}$$

with the periodic boundary conditions

$$\phi_0 = \phi_I, \qquad \phi_{I+1} = \phi_1, \tag{37}$$

where $\phi_1, \phi_2, \ldots, \phi_I$ are unknown, whereas a_i, b_i, c_i and F_i are given parameters. The system can be written as

$$A\vec{\phi} = \vec{F}, \tag{38}$$

where $\vec{\phi} = (\phi_1, \phi_2, \ldots, \phi_I)^T$, $\vec{F} = (F_1, F_2, \ldots, F_I)^T$ and

$$A = \begin{bmatrix} -b_1 & c_1 & 0 & 0 & a_1 \\ a_2 & -b_2 & c_2 & 0 & 0 \\ & \ddots & \ddots & \ddots & \\ 0 & 0 & a_{I-1} & -b_{I-1} & c_{I-1} \\ c_I & 0 & 0 & a_I & -b_I \end{bmatrix}. \tag{39}$$

A non-iterative method for solving problem (38) is based on the Sherman-Morrison formula [16, 17]. Indeed, the matrix

$$A = C + \vec{u}\vec{v}^T \tag{40}$$

differs from the tridiagonal matrix by only two elements — $a_1 \neq 0$ and $c_I \neq 0$; here $\vec{u} = (1, 0, \ldots, c_I)^T$, $\vec{v}^T = (1, 0, \ldots, a_1)$. Hence, the solution to system (38) is

$$\vec{\phi} = (C + \vec{u}\vec{v}^T)^{-1}\vec{F} = C^{-1}\vec{F} - \alpha^{-1}(C^{-1}\vec{u})\vec{v}^T(C^{-1}\vec{F}) = \vec{x} - \alpha^{-1}(\vec{v}^T\vec{x})\vec{z}, \tag{41}$$

where

$$\vec{x} = C^{-1}\vec{F}, \qquad \vec{z} = C^{-1}\vec{u} \tag{42}$$

and $\alpha^{-1} = 1 + \vec{v}^T\vec{z}$. Since C is a tridiagonal matrix, both systems

$$C\vec{x} = \vec{F}, \qquad C\vec{z} = \vec{u} \tag{43}$$

are solved using the direct Thomas factorisation method [18]. Then α^{-1} is calculated, and the solution follows from (41).

3.3.2. Solution of the Split Systems in the ϑ-Direction

We now develop a direct (i.e., non-iterative) method for the solution of the second and the fourth 1D equations of system (29) in the ϑ-direction. The method is based on bordering the block structure of the system matrix. For each fixed index $i = \overline{1, I}$ one has to solve the following system of equations

$$\begin{aligned}
-\bar{b}\phi_0 + \sum_{i=1}^{I} \bar{c}_i\phi_{i,1} &= -F_0, \qquad j = 0 \text{ (North pole)}, \\
a_{ij}\phi_{i,j-1} - b_{ij}\phi_{ij} + c_{ij}\phi_{i,j+1} &= -F_{ij}, \qquad j = \overline{1, J}, \\
-\tilde{b}\phi_{J+1} + \sum_{i=1}^{I} \tilde{a}_i\phi_{i,J} &= -F_{J+1}, \qquad j = J+1 \text{ (South pole)},
\end{aligned} \tag{44}$$

where $\vec{\phi} = \{\phi_0, \phi_{i,1}, \ldots, \phi_{i,J}, \phi_{J+1}\}^T$ is the unknown vector, while a_{ij}, b_{ij}, c_{ij}, F_{ij}, $\tilde{a}_i$, $\bar{b}$, $\tilde{b}$, $\bar{c}_i$, F_0 and F_{J+1} are given parameters. This system can be written as

$$A\vec{\phi} = -\vec{F}, \tag{45}$$

where

$$\vec{\phi} = \begin{bmatrix} \phi_0 \\ \vec{\phi}_1 \\ \vdots \\ \vec{\phi}_J \\ \phi_{J+1} \end{bmatrix}, \vec{F} = \begin{bmatrix} F_0 \\ \vec{F}_1 \\ \vdots \\ \vec{F}_J \\ F_{J+1} \end{bmatrix}, \vec{\phi}_i = \begin{bmatrix} \phi_{i,1} \\ \phi_{i,2} \\ \vdots \\ \phi_{i,J-1} \\ \phi_{i,J} \end{bmatrix}, \vec{F}_i = \begin{bmatrix} F_{i,1} \\ F_{i,2} \\ \vdots \\ F_{i,J-1} \\ F_{i,J} \end{bmatrix}, \qquad i = \overline{1, I},$$

whereas

$$A = \begin{bmatrix} -\bar{b} & \vec{X}_1 & \vec{X}_2 & \dots & \vec{X}_I & 0 \\ \vec{Y}_1 & P_1 & 0 & \dots & 0 & \vec{W}_1 \\ \vec{Y}_2 & 0 & P_2 & \dots & 0 & \vec{W}_2 \\ \dots & \dots & \dots & \dots & \dots & \dots \\ \vec{Y}_I & 0 & 0 & \dots & P_I & \vec{W}_I \\ 0 & \vec{U}_1 & \vec{U}_2 & \dots & \vec{U}_I & -\tilde{b} \end{bmatrix} \tag{46}$$

is a block (partitioned) matrix with the vector elements of dimension J

$$\begin{aligned} \vec{X}_i = (\bar{c}_i, 0, \dots, 0), \qquad \vec{U}_i = (0, \dots, 0, \tilde{a}_i), \\ \vec{Y}_i^T = (a_{i,1}, 0, \dots, 0), \qquad \vec{W}_i^T = (0, \dots, 0, c_{i,J}) \end{aligned} \tag{47}$$

and the tridiagonal matrix elements P_i of dimension $J \times J$. The block structure of matrix (46) allows us to represent the solution as $\vec{\phi} = \{\phi_0, \vec{\phi}_i^T, \phi_{J+1}\}^T$, where $\vec{\phi}_i^T = \{\phi_{i,1}, \dots, \vec{\phi}_{i.J}\}$, and to apply the bordering method for non-iterative solving system (44). In fact, then system (44) can be written as

$$\begin{aligned} -\bar{b}\phi_0 + \sum_{i=1}^{I} \langle \vec{X}_i, \vec{\phi}_i \rangle &= -F_0, \\ \phi_0 \vec{Y}_i + P_i \vec{\phi}_i + \phi_{J+1} \vec{W}_i &= -\vec{F}_i, \qquad i = \overline{1, I}, \\ \sum_{i=1}^{I} \langle \vec{U}_i, \vec{\phi}_i \rangle - \tilde{b}\phi_{J+1} &= -F_{J+1}. \end{aligned} \tag{48}$$

We search for the solution in the form

$$\vec{\phi}_i = \vec{G}_i + \phi_0 \vec{V}_i + \phi_{J+1} \vec{Z}_i, \qquad i = \overline{1, I}, \tag{49}$$

where $\vec{G}_i$, $\vec{V}_i$ and $\vec{Z}_i$ are the solutions to the following three systems

$$P_i\vec{G}_i = -\vec{F}_i, \qquad P_i\vec{V}_i = -\vec{Y}_i, \qquad P_i\vec{Z}_i = -\vec{W}_i, \qquad i = \overline{1, I}, \tag{50}$$

respectively. Since each P_i is a tridiagonal matrix, the Thomas factorisation method can be used again for the solution of systems (50). Moreover, the problems (50) can be solved simultaneously by using parallel processors. Note that for each i only one inverse matrix P_i^{-1} is required to solve all the three problems (50).

Once the vectors $\vec{G}_i$, $\vec{V}_i$ and $\vec{Z}_i$ are determined, the solution values ϕ_0 and ϕ_{J+1} at the pole points are calculated as

$$\phi_0 = R_0/R, \qquad \phi_{J+1} = R_{J+1}/R, \tag{51}$$

where

$$R_0 = \left(\sum_{i=1}^{I}\langle\vec{U}_i, \vec{Z}_i\rangle - \tilde{b}\right)\left(F_0 + \sum_{i=1}^{I}\langle\vec{X}_i, \vec{G}_i\rangle\right) - \left(\sum_{i=1}^{I}\langle\vec{X}_i, \vec{Z}_i\rangle\right)\left(F_{J+1} + \sum_{i=1}^{I}\langle\vec{U}_i, \vec{G}_i\rangle\right), \tag{52}$$

$$R_{J+1} = \left(\sum_{i=1}^{I}\langle\vec{X}_i, \vec{V}_i\rangle - \bar{b}\right)\left(F_{J+1} + \sum_{i=1}^{I}\langle\vec{U}_i, \vec{G}_i\rangle\right) - \left(\sum_{i=1}^{I}\langle\vec{U}_i, \vec{V}_i\rangle\right)\left(F_0 + \sum_{i=1}^{I}\langle\vec{X}_i, \vec{G}_i\rangle\right), \tag{53}$$

$$R = \left(\sum_{i=1}^{I}\langle\vec{X}_i, \vec{Z}_i\rangle\right)\left(\sum_{i=1}^{I}\langle\vec{U}_i, \vec{V}_i\rangle\right) - \left(\sum_{i=1}^{I}\langle\vec{X}_i, \vec{V}_i\rangle - \bar{b}\right)\left(\sum_{i=1}^{I}\langle\vec{U}_i, \vec{Z}_i\rangle - \tilde{b}\right). \tag{54}$$

4. General Diffusion Problem

In the particular case of $\mathbf{U} = \mathbf{0}$ and $\sigma = 0$ the advection-diffusion equation (1) reduces to the linear diffusion problem

$$\frac{\partial T}{\partial t} = AT + f, \qquad T(\lambda, \varphi, 0) = g(\lambda, \varphi), \tag{55}$$

where A is the diffusion operator

$$AT =\equiv \frac{1}{R\cos\varphi}\left[\frac{\partial}{\partial\lambda}\left(\frac{\mu}{R\cos\varphi}\frac{\partial T}{\partial\lambda}\right) + \frac{\partial}{\partial\varphi}\left(\frac{\mu\cos\varphi}{R}\frac{\partial T}{\partial\varphi}\right)\right]. \tag{56}$$

Here $T(\lambda, \varphi, t) \geq 0$ is, depending on the application, the density of a substance, the temperature, etc., $\mu = \mu(\lambda, \varphi) \geq 0$ is the diffusion coefficient, $f = f(\lambda, \varphi) \geq 0$ is the source function, R is the radius of the sphere S, λ is the longitude (positive eastward) and φ is the latitude (positive northward).

Integrating (55) over the domain S, we find that the terms containing the spatial derivatives vanish, and so

$$\frac{d}{dt}\int_S T dS = \int_S f dS. \tag{57}$$

Expression (57) is called the balance equation [19]. If both sides of (55), prior to the integration, are multiplied by T then we shall obtain

$$\frac{1}{2}\frac{d}{dt}\int_S T^2 dS = \int_S TAT dS + \int_S fT dS. \tag{58}$$

The integral on the left-hand side of (58) is often called the energy of solution (in the L_2-norm). Because $\int_S TATdS \leq 0$, the diffusion operator A is negative definite, and so the first summand on the right-hand side of (58) serves as the sink of energy, while the second summand serves as the source thereof. Consequently,

$$\frac{1}{2}\frac{d}{dt}\int_S T^2 dS \leq \int_S fT dS. \tag{59}$$

In the particular case of $f = 0$ the balance equation (57) turns into the mass conservation law

$$\frac{d}{dt}\int_S T dS = 0, \tag{60}$$

while inequality (59) yields the solution's dissipation in the L_2-norm

$$\frac{1}{2}\frac{d}{dt}\int_S T^2 dS \leq 0. \tag{61}$$

4.1. Analytical Solution of the Linear Diffusion Problem

Consider the diffusion equation (55) with constant coefficient μ. Then $AT \equiv \mu\Delta T$, where Δ is the spherical Laplace operator, so that the diffusion problem can be rewritten as

$$\frac{\partial T}{\partial t} = \mu \Delta T + f, \qquad T(\lambda, \varphi, 0) = g(\lambda, \varphi). \tag{62}$$

Let

$$\langle g_1, g_2 \rangle = \int_S g_1(\lambda, \varphi, t)\overline{g_2(\lambda, \varphi, t)}dS \tag{63}$$

be the inner product in the Hilbert space $L_2(S)$. Here $\overline{g_2(\lambda, \varphi, t)}$ denotes the complex conjugate to g_2. It is known [43] that spherical harmonics

$$Y_n^m(\lambda, \varphi) = \left[\frac{2n+1}{4\pi}\frac{(n-m)!}{(n+m)!}\right]^{1/2} P_n^m(\varphi)\exp(im\lambda), n \geq 0, |m| \leq n \tag{64}$$

form an orthonomal basis in $L_2(S)$, that is

$$\left\langle Y_n^m, Y_l^k \right\rangle = \delta_{mk}\delta_{nl}, \tag{65}$$

where δ_{mk} and δ_{nl} are the Kronecker delta's, whereas $P_n^m(\varphi)$ is the associated Legendre function of degree n and zonal wavenumber m. Let n and m be integer, $n \geq 0$ and $|m| \leq n$. Each spherical harmonic $Y_n^m(\lambda, \varphi)$ is an eigenfunction of the spectral problem

$$\Delta Y_n^m = -n(n+1)Y_n^m, \qquad |m| \leq n, \tag{66}$$

corresponding to the eigenvalue $\chi_n = -n(n+1)$ of multiplicity $2n+1$. Using the Fourier-Laplace series

$$\begin{aligned} T(\lambda, \varphi, t) &= \sum_{n=1}^{\infty}\sum_{m=-n}^{n} T_n^m(t)Y_n^m(\lambda, \varphi), \\ f(\lambda, \varphi, t) &= \sum_{n=1}^{\infty}\sum_{m=-n}^{n} f_n^m(t)Y_n^m(\lambda, \varphi), \\ g(\lambda, \varphi) &= \sum_{n=1}^{\infty}\sum_{m=-n}^{n} g_n^m Y_n^m(\lambda, \varphi), \end{aligned} \tag{67}$$

(all the functions are assumed to be orthogonal to a constant on S), calculating the inner product of (62) with each of the spherical harmonics Y_l^k, given (66) we obtain

$$\frac{dT_l^k}{dt} = -\mu l\,(l+1)\,T_l^k + f_l^k, \qquad T_l^k(0) = g_l^k, \qquad l \geq 1, \qquad |k| \leq l. \quad (68)$$

Integration of (68) from $t = 0$ to t yields

$$T_l^k(t) = g_l^k \exp(-\mu l\,(l+1)\,t) + \int_0^t f_l^k(t)dt, \qquad l \geq 1, \qquad |k| \leq l. \quad (69)$$

Hence, the first formula in (67) provides the solution. In the case of $f(\lambda, \varphi, t) \equiv 0$ formula (69) reduces to

$$T_l^k(t) = g_l^k \exp(-\mu l\,(l+1)\,t), \qquad l \geq 1, \qquad |k| \leq l. \quad (70)$$

In the most general case the diffusion coefficient μ is variable. If so, one has to use numerical methods for solving the diffusion problem.

There exists voluminous literature on numerical solution of the diffusion equation in a multidimensional cube [41, 44–54]. In particular, the approach suggested by Gibou and Fedkiw [55] can be employed for constructing a five-point fourth-order finite difference scheme in a doubly periodic domain.

A novel numerical method for solving the diffusion problem on the sphere, based on the idea of splitting, is presented below.

5. Grid-Swap Methods (Methods 2 and 3)

5.1. Operator Splitting

Define the grid spacings in the standard way: $\tau = t_{n+1} - t_n$, $\Delta\lambda = \lambda_{k+1} - \lambda_k$, $\Delta\varphi = \varphi_{l+1} - \varphi_l$. Then in every (sufficiently small) time interval (t_n, t_{n+1}) we can split (55) by coordinates as follows

$$\frac{\partial T}{\partial t} = A_\lambda T + \frac{f}{2} = \frac{1}{R\cos\varphi}\frac{\partial}{\partial\lambda}\left(\frac{\mu}{R\cos\varphi}\frac{\partial T}{\partial\lambda}\right) + \frac{f}{2}, \quad (71)$$

$$\frac{\partial T}{\partial t} = A_\varphi T + \frac{f}{2} = \frac{1}{R\cos\varphi}\frac{\partial}{\partial\varphi}\left(\frac{\mu\cos\varphi}{R}\frac{\partial T}{\partial\varphi}\right) + \frac{f}{2}. \quad (72)$$

The splitting implies that equations (71)-(72) are being solved successively, one after another, so that the solution to the previous equation serves as the initial condition for the current one. Besides, as we split the longitudinal and latitudinal terms, the latitude is fixed while solving (71), as well as the longitude is fixed when solving (72) [40,41,46,50,53].

5.2. Two Coordinate Maps of the Sphere

Because the denominator of the metric term $(R\cos\varphi)^{-1}$ vanishes in the poles of the sphere, we exclude the poles defining the grid on S as follows (Figure 2)

$$S^{(1)}_{\Delta\lambda,\Delta\varphi} = \left\{(\lambda_k,\varphi_l) : \lambda_k \in \left[\tfrac{\Delta\lambda}{2}, 2\pi + \tfrac{\Delta\lambda}{2}\right), \varphi_l \in \left[-\tfrac{\pi}{2} + \tfrac{\Delta\varphi}{2}, \tfrac{\pi}{2} - \tfrac{\Delta\varphi}{2}\right]\right\}. \quad (73)$$

Such a shift of the grid a half step in φ allows the diffusion equation to have sense everywhere on the discrete sphere $S^{(1)}_{\Delta\lambda,\Delta\varphi}$, which can approximate S as accurate as possible. Moreover, because of the splitting we can change the coordinate map from (73) to

$$S^{(2)}_{\Delta\lambda,\Delta\varphi} = \left\{(\lambda_k,\varphi_l) : \lambda_k \in \left[\tfrac{\Delta\lambda}{2}, \pi - \tfrac{\Delta\lambda}{2}\right], \varphi_l \in \left[-\tfrac{\pi}{2} + \tfrac{\Delta\varphi}{2}, \tfrac{3\pi}{2} + \tfrac{\Delta\varphi}{2}\right)\right\}. \quad (74)$$

Grid (73) is used for computing the solution in λ, while grid (74) serves for computing in φ. Obviously, both grids are defined on the same nodes (λ_k, φ_l) (Figure 3). An essential benefit of the use of the two grids is that we can involve simple periodic boundary conditions in both directions, thereby avoiding many undesired numerical procedures such as constructing artificial boundary conditions, performing matrix bordering, joining the solution from two opposite meridians at the poles, etc. This circumstance is particularly remarkable since sphere is *not* a doubly periodic domain.

5.3. Second-Order Finite Difference Schemes (Method 2)

Having involved the two coordinate grids, now we approximate the temporal and spatial derivatives in (71)-(72). For the temporal derivatives in (71) and (72) we take

$$\left.\frac{\partial T}{\partial t}\right|_{t=t_n} \approx \frac{T^{n+\frac{1}{2}}_{kl} - T^{n}_{kl}}{\tau}, \qquad \left.\frac{\partial T}{\partial t}\right|_{t=t_{n+\frac{1}{2}}} \approx \frac{T^{n+1}_{kl} - T^{n+\frac{1}{2}}_{kl}}{\tau}, \quad (75)$$

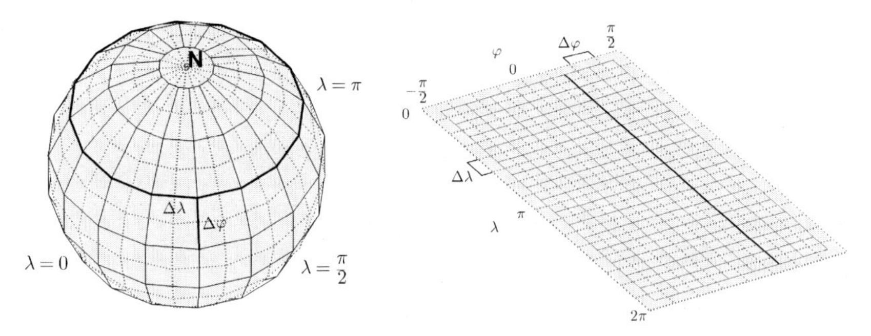

Figure 2. Representation of the sphere in the grid-swap methods: grid $S^{(1)}_{\Delta\lambda,\Delta\varphi}$ (solid lines) is used while computing in λ.

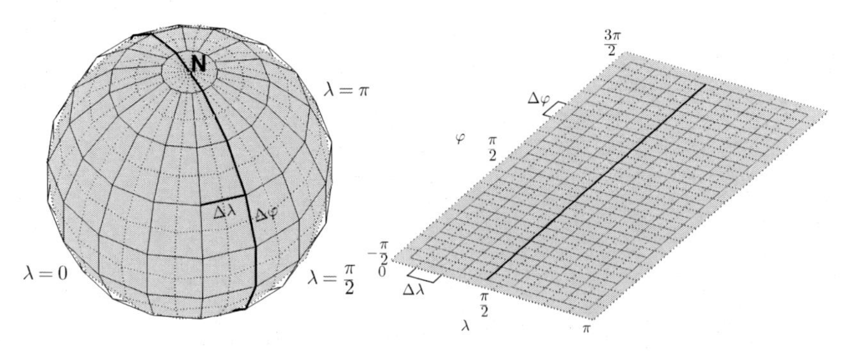

Figure 3. Representation of the sphere in the grid-swap methods: grid $S^{(2)}_{\Delta\lambda,\Delta\varphi}$ (solid lines) is used while computing in φ.

respectively. And for the spatial derivatives in (71) and (72) we take

$$\frac{\partial}{\partial\lambda}\left(\mu\frac{\partial T}{\partial\lambda}\right)\Bigg|_{\lambda=\lambda_k} \approx \frac{1}{\Delta\lambda}\left(\mu_{k+1/2}\frac{T_{k+1}-T_k}{\Delta\lambda} - \mu_{k-1/2}\frac{T_k - T_{k-1}}{\Delta\lambda}\right) \quad (76)$$

and

$$\frac{\partial}{\partial\varphi}\left(\mu|\cos\varphi|\frac{\partial T}{\partial\varphi}\right)\Bigg|_{\varphi=\varphi_l} \approx \frac{1}{\Delta\varphi}\left[(\mu|\cos\varphi|)_{l+\frac{1}{2}}\frac{T_{l+1}-T_l}{\Delta\varphi} - (\mu|\cos\varphi|)_{l-\frac{1}{2}}\frac{T_l - T_{l-1}}{\Delta\varphi}\right], \quad (77)$$

where for the sake of clarity the corresponding fixed indexes l and k are omitted, $|\cos\varphi|$ appears since $\varphi \in \left[-\frac{\pi}{2} + \frac{\Delta\varphi}{2}, \frac{3\pi}{2} + \frac{\Delta\varphi}{2}\right)$ in (74), whereas

$$\mu_{k+p/2} := \frac{\mu_{k+(p+1)/2} + \mu_{k+(p-1)/2}}{2}, p = \pm 1, \quad (78)$$

$$(\mu|\cos\varphi|)_{l+p/2} := \frac{(\mu|\cos\varphi|)_{l+(p+1)/2} + (\mu|\cos\varphi|)_{l+(p-1)/2}}{2}, p = \pm 1. \quad (79)$$

Lemma 5 *[9] Let grids (73) and (74) take place and hence the solutions of one-dimensional equations (71) and (72) can be treated as periodic, both in λ and φ. Then in Method 2 the discrete diffusion operators in (76)-(77) are negative definite (energy dissipative in the $L_2(S^{(1)}_{\Delta\lambda,\Delta\varphi})$- and $L_2(S^{(2)}_{\Delta\lambda,\Delta\varphi})$-norms, respectively), because*

$$\sum_j \left[\gamma_{j+\frac{1}{2}}(T_{j+1} - T_j) - \gamma_{j-\frac{1}{2}}(T_j - T_{j-1})\right] T_j = -\sum_j \gamma_{j+\frac{1}{2}}(T_{j+1} - T_j)^2 \leq 0.$$

For the temporal discretisation of the spatial terms we use the Crank-Nicolson approximation

$$T \approx \frac{T^{n+\frac{p}{2}} + T^{n+\frac{p-1}{2}}}{2}, \quad (80)$$

where $p = 1$ for (71) and $p = 2$ for (72), and $f = f^{n+\frac{1}{2}} = f(\lambda, \varphi, t_n + \tau/2)$ in (71) and (72) [40, 56]. Therefore, the second-order finite difference scheme has the following equations:
— in the λ-direction

$$-T^{n+\frac{1}{2}}_{k+1} m_{k+1} + T^{n+\frac{1}{2}}_{k}\left(\tau^{-1} + m_k\right) - T^{n+\frac{1}{2}}_{k-1} m_{k-1} =$$
$$T^n_{k+1} m_{k+1} + T^n_k \left(\tau^{-1} - m_k\right) + T^n_{k-1} m_{k-1} + \frac{f^{n+\frac{1}{2}}_k}{2}, \quad (81)$$

where

$$m_k = \frac{\mu_{k+1/2} + \mu_{k-1/2}}{2\Delta\lambda^2 R^2 \cos^2\varphi_l}, \qquad m_{k+p} = \frac{\mu_{k+p/2}}{2\Delta\lambda^2 R^2 \cos^2\varphi_l}, \qquad p = \pm 1; \quad (82)$$

— in the φ-direction

$$-T_{l+1}^{n+1}m_{l+1}+T_l^{n+1}\left(\tau^{-1}+m_l\right)-T_{l-1}^{n+1}m_{l-1}=$$
$$T_{l+1}^{n+\frac{1}{2}}m_{l+1}+T_l^{n+\frac{1}{2}}\left(\tau^{-1}-m_l\right)+T_{l-1}^{n+\frac{1}{2}}m_{l-1}+\frac{f_l^{n+\frac{1}{2}}}{2}, \tag{83}$$

where

$$m_l=\frac{(\mu|\cos\varphi|)_{l+1/2}+(\mu|\cos\varphi|)_{l-1/2}}{2R^2|\cos\varphi_l|\Delta\varphi^2},\ m_{l+p}=\frac{(\mu|\cos\varphi|)_{l+p/2}}{2R^2|\cos\varphi_l|\Delta\varphi^2}, p=\pm 1. \tag{84}$$

Due to Lemma 5 the resulting matrices in (81) and (83) are positive definite.

Theorem 2 *Let τ be such that $(1-\tau m_k)>0$ for all k and $(1-\tau m_l)>0$ for all l. Then schemes (81) and (83) are monotonic.*

Proof. It is sufficient to prove the monotonicity of (81). System (81) can be written as $A\vec{T}=\vec{b}$, where the matrix A is positive definite as a symmetric strictly diagonally dominant real matrix with positive diagonal entries [57]. Since its non-diagonal elements are negative, A is an M-matrix (see [57, sec. 36.16]), and hence, it is monotonic (see [57, sec. 36.9]). Thus, the condition $\vec{b}=A\vec{T}>0$ implies $\vec{T}>0$ [58], and scheme (81) is monotonic. ■

5.4. Fourth-Order Finite Difference Schemes (Method 3)

Since we involve periodic boundary conditions in both directions, increase of the approximation order in space can be used to develop a fourth-order finite difference scheme [37–39]. For this we represent the spatial terms in (71)-(72) as

$$\frac{\partial}{\partial\lambda}\left(\mu\frac{\partial T}{\partial\lambda}\right)=\frac{\partial\mu}{\partial\lambda}\frac{\partial T}{\partial\lambda}+\mu\frac{\partial^2 T}{\partial\lambda^2}, \tag{85}$$

$$\frac{1}{\cos\varphi}\frac{\partial}{\partial\varphi}\left(\mu\cos\varphi\frac{\partial T}{\partial\varphi}\right)=\left(\frac{\partial\mu}{\partial\varphi}-\mu\tan\varphi\right)\frac{\partial T}{\partial\varphi}+\mu\frac{\partial^2 T}{\partial\varphi^2} \tag{86}$$

and then approximate (85) and (86) using the fourth-order central stencils as follows (e.g., [55])

$$\left.\frac{\partial \mu}{\partial \lambda}\frac{\partial T}{\partial \lambda}+\mu\frac{\partial^2 T}{\partial \lambda^2}\right|_{\lambda=\lambda_k} \approx$$
$$\left(\frac{-\mu_{k+2}+8\mu_{k+1}-8\mu_{k-1}+\mu_{k-2}}{12\Delta\lambda}\right)\left(\frac{-T_{k+2}+8T_{k+1}-8T_{k-1}+T_{k-2}}{12\Delta\lambda}\right) \quad (87)$$
$$+\mu_k\frac{-T_{k+2}+16T_{k+1}-30T_k+16T_{k-1}-T_{k-2}}{12\Delta\lambda^2},$$

$$\left.\left(\frac{\partial \mu}{\partial \varphi}-\mu\tan\varphi\right)\frac{\partial T}{\partial \varphi}+\mu\frac{\partial^2 T}{\partial \varphi^2}\right|_{\varphi=\varphi_l} \approx$$
$$\left(\frac{-\mu_{l+2}+8\mu_{l+1}-8\mu_{l-1}+\mu_{l-2}}{12\Delta\varphi}-\mu_l\tan\varphi_l\right)\frac{-T_{l+2}+8T_{l+1}-8T_{l-1}+T_{l-2}}{12\Delta\varphi}$$
$$+\mu_l\frac{-T_{l+2}+16T_{l+1}-30T_l+16T_{l-1}-T_{l-2}}{12\Delta\varphi^2}. \quad (88)$$

Then the fourth-order finite difference scheme has the following equations:
— in the λ-direction

$$T_{k+2}^{n+\frac{1}{2}}m_{k+2}-T_{k+1}^{n+\frac{1}{2}}m_{k+1}+T_k^{n+\frac{1}{2}}\left(\tau^{-1}+m_k\right)-T_{k-1}^{n+\frac{1}{2}}m_{k-1}+T_{k-2}^{n+\frac{1}{2}}m_{k-2}=$$
$$-T_{k+2}^n m_{k+2}+T_{k+1}^n m_{k+1}+T_k^n\left(\tau^{-1}-m_k\right)+T_{k-1}^n m_{k-1}-T_{k-2}^n m_{k-2}+\frac{f_k^{n+\frac{1}{2}}}{2}, \quad (89)$$

where

$$m_k = \frac{30\mu_k}{24\Delta\lambda^2R^2\cos^2\varphi_l}, \qquad M_k=\frac{-\mu_{k+2}+8\mu_{k+1}-8\mu_{k-1}+\mu_{k-2}}{12\Delta\lambda},$$
$$m_{k+p} = \frac{1}{R^2\cos^2\varphi_l}\left(\frac{\mu_k}{24\Delta\lambda^2}+\mathrm{sgn}(p)\frac{M_k}{24\Delta\lambda}\right), \qquad p=\pm 2,$$
$$m_{k+p} = \frac{1}{R^2\cos^2\varphi_l}\left(\frac{16\mu_k}{24\Delta\lambda^2}+\mathrm{sgn}(p)\frac{8M_k}{24\Delta\lambda}\right), \qquad p=\pm 1; \quad (90)$$

— in the φ-direction

$$T_{l+2}^{n+1}m_{l+2}-T_{l+1}^{n+1}m_{l+1}+T_l^{n+1}\left(\tau^{-1}+m_l\right)-T_{l-1}^{n+1}m_{l-1}+T_{l-2}^{n+1}m_{l-2}=$$
$$-T_{l+2}^{n+\frac{1}{2}}m_{l+2}+T_{l+1}^{n+\frac{1}{2}}m_{l+1}+T_l^{n+\frac{1}{2}}\left(\tau^{-1}-m_l\right)+T_{l-1}^{n+\frac{1}{2}}m_{l-1}-T_{l-2}^{n+\frac{1}{2}}m_{l-2}+\frac{f_l^{n+\frac{1}{2}}}{2}, \quad (91)$$

where

$$
\begin{aligned}
m_l &= \frac{30\mu_l}{24R^2\Delta\varphi^2}, \qquad M_l = \frac{-\mu_{l+2}+8\mu_{l+1}-8\mu_{l-1}+\mu_{l-2}}{12\Delta\varphi} - \mu_l \tan\varphi_l, \\
m_{l+p} &= \frac{1}{R^2}\left(\frac{\mu_l}{24\Delta\varphi^2} + \mathrm{sgn}(p)\frac{M_l}{24\Delta\varphi}\right), \qquad p = \pm 2, \\
m_{l+p} &= \frac{1}{R^2}\left(\frac{16\mu_l}{24\Delta\varphi^2} + \mathrm{sgn}(p)\frac{8M_l}{24\Delta\varphi}\right), \qquad p = \pm 1. \qquad (92)
\end{aligned}
$$

Lemma 6 *[9] Let grids (73) and (74) take place and let the diffusion coefficient μ be constant. Then in Method 3 the fourth-order finite difference diffusion operators (87)-(88) are negative definite, because*

$$
\sum_k \left[-T_{k+2} + 16T_{k+1} - 30T_k + 16T_{k-1} - T_{k-2}\right] T_k \leq -12\sum_k (T_{k+1} - T_k)^2 \leq 0.
$$

Remark 1 *Lemma 6 establishes negative definiteness of the operators in (87)-(88) only for a constant diffusion coefficient. In all the numerical experiments considered below, in which μ was variable, the results also showed negative definiteness of the corresponding finite difference operators.*

Schemes (81)-(84) and (89)-(92) have the second and fourth orders of approximation in space, respectively, but of the first order in time. Employing the Strang splitting [59] (also known as dicyclic splitting [40])

$$
\begin{aligned}
\frac{\vec{T}^{n+1/5} - \vec{T}^n}{\tau/2} &= A_{\Delta\lambda}\frac{\vec{T}^{n+1/5} + \vec{T}^n}{2}, \\
\frac{\vec{T}^{n+2/5} - \vec{T}^{n+1/5}}{\tau/2} &= A_{\Delta\varphi}\frac{\vec{T}^{n+2/5} + \vec{T}^{n+1/5}}{2}, \\
\frac{\vec{T}^{n+3/5} - \vec{T}^{n+2/5}}{\tau} &= \vec{f}^{n+1/2}, \qquad (93) \\
\frac{\vec{T}^{n+4/5} - \vec{T}^{n+3/5}}{\tau/2} &= A_{\Delta\varphi}\frac{\vec{T}^{n+4/5} + \vec{T}^{n+3/5}}{2}, \\
\frac{\vec{T}^{n+1} - \vec{T}^{n+4/5}}{\tau/2} &= A_{\Delta\lambda}\frac{\vec{T}^{n+1} + \vec{T}^{n+4/5}}{2},
\end{aligned}
$$

leads to a scheme of second order accuracy in time in the interval (t_n, t_{n+1}) [40, 41, 59] (also [60]). Here $\vec{T}^p$ and $\vec{f}^{n+1/2}$ are the vectors with components $\{T_{kl}^p\}$ and $\{f_{kl}^{n+1/2}\}$, while $A_{\Delta\lambda}$ and $A_{\Delta\varphi}$ are the finite difference operators (matrices) approximating the differential operators A_λ and A_φ of the split problems (71) and (72) in the directions of λ and φ, respectively.

Theorem 3 *Let (73) and (74) take place, let μ be arbitrary in Method 2 and constant in Method 3. Then the whole 2D split problem (93) appearing under Methods 2 and 3 satisfies the discrete analogue of the balance equation (57)*

$$\sum_k \sum_l \left[T_{kl}^{n+1} - T_{kl}^n\right] \Delta S_{kl} = \tau \sum_k \sum_l f_{kl}^{n+1/2} \Delta S_{kl}, \tag{94}$$

where $\Delta S_{kl} = \Delta\lambda\Delta\varphi \cos\varphi$.

The proof is trivial.

Corollary 3 *If $f = 0$ then (94) turns into the discrete form of the mass conservation law (60)*

$$\sum_k \sum_l T_{kl}^{n+1} \Delta S_{kl} = \sum_k \sum_l T_{kl}^n \Delta S_{kl}. \tag{95}$$

Theorem 4 *Let μ be arbitrary in Method 2 and constant in Method 3. Then the whole 2D split problem (93) appearing under Methods 2 and 3 is unconditionally stable*

$$\|\vec{T}^{n+1}\| \leq \|\vec{T}^n\| + \tau \|\vec{f}^{n+1/2}\|, \tag{96}$$

where

$$\|\vec{\phi}^n\| = \left\langle \vec{\phi}^n, \vec{\phi}^n \right\rangle^{\frac{1}{2}} = \left(\sum_k \sum_l |\phi_{kl}^n|^2 \Delta S_{kl} \right)^{\frac{1}{2}}. \tag{97}$$

Proof. To prove Theorem 4, it is sufficient to rewrite scheme (93) in the equivalent form

$$\vec{T}^{n+1} = A_1 A_2 A_2 A_1 \vec{T}^n + \tau A_1 A_2 \vec{f}^{n+1/2},$$

where $A_1 = \left(E - \frac{1}{2}\tau A_{\Delta\lambda}\right)^{-1} \left(E + \frac{1}{2}\tau A_{\Delta\lambda}\right)$, $A_2 = \left(E - \frac{1}{2}\tau A_{\Delta\varphi}\right)^{-1} \left(E + \frac{1}{2}\tau A_{\Delta\varphi}\right)$ and E is the unit matrix. Then (96) follows from the inequalities $\|A_i\| \leq 1$, $i = 1, 2$, valid for the spectral matrix norm [40]. ■

Corollary 4 *If $f = 0$ then (96) demonstrates dissipation, i.e., $\|\vec{T}^{n+1}\| \le \|\vec{T}^{n}\|$.*

It is important that the constructed finite difference schemes are systems of linear algebraic equations with band matrices, and hence the numerical solution can be found by fast direct algorithms (e.g., in the case of Method 2 (second-order schemes), the matrices are tridiagonal, so the Sherman-Morrison formula and the Thomas algorithm are used [18,61]).

5.5. Methods 2 and 3: Nonlinear Diffusion Equation

The nonlinear diffusion equation on the sphere S has the form

$$\frac{\partial T}{\partial t} = AT + f, \tag{98}$$

where

$$AT \equiv \frac{1}{R\cos\varphi}\left[\frac{\partial}{\partial\lambda}\left(\frac{\mu T^{\alpha}}{R\cos\varphi}\frac{\partial T}{\partial\lambda}\right) + \frac{\partial}{\partial\varphi}\left(\frac{\mu T^{\alpha}\cos\varphi}{R}\frac{\partial T}{\partial\varphi}\right)\right], \tag{99}$$

Here, unlike the linear diffusion equation (55), the diffusion coefficient additionally depends on the solution itself, where the parameter α is (usually) a positive integer number that determines the degree of nonlinearity of the diffusion process; the particular case $\alpha = 0$ corresponds to the linear diffusion problem.

To deal with the nonlinear problem, in every time interval (t_n, t_{n+1}) we first linearise (98) and then split the linearised equation by coordinates [3,54]

$$\frac{\partial T}{\partial t} = \frac{1}{R\cos\varphi}\frac{\partial}{\partial\lambda}\left(\frac{D}{R\cos\varphi}\frac{\partial T}{\partial\lambda}\right) + \frac{f}{2}, \tag{100}$$

$$\frac{\partial T}{\partial t} = \frac{1}{R\cos\varphi}\frac{\partial}{\partial\varphi}\left(\frac{D\cos\varphi}{R}\frac{\partial T}{\partial\varphi}\right) + \frac{f}{2}, \tag{101}$$

where

$$D = \mu(T^n)^{\alpha}, \qquad T^n = T(\lambda, \varphi, t_n). \tag{102}$$

Applying to (100)-(101) the same procedure that we involved for the linear diffusion problem, we shall finally obtain the second-order finite difference scheme:

— in the λ-direction we have equation (81) again, but now

$$m_k = \frac{D_{k+1/2} + D_{k-1/2}}{2R^2 \Delta\lambda^2 \cos^2 \varphi_l}, \qquad m_{k+p} = \frac{D_{k+p/2}}{2R^2 \Delta\lambda^2 \cos^2 \varphi_l}, \qquad p \pm 1; \quad (103)$$

— in the φ-direction we have equation (83) again, but now

$$m_l = \frac{(D|\cos\varphi|)_{l+1/2} + (D|\cos\varphi|)_{l-1/2}}{2R^2 |\cos\varphi_l| \Delta\varphi^2}, m_{l+p} = \frac{(D|\cos\varphi|)_{l+p/2}}{2R^2 |\cos\varphi_l| \Delta\varphi^2}, p = \pm 1. \quad (104)$$

For the fourth-order scheme we shall obtain:

— in the λ-direction we have equation (89), where

$$\begin{aligned} m_k &= \frac{30 D_k}{24 \Delta\lambda^2 R^2 \cos^2 \varphi_l}, M_k = \frac{-D_{k+2} + 8D_{k+1} - 8D_{k-1} + D_{k-2}}{12\Delta\lambda}, \\ m_{k+p} &= \frac{1}{R^2 \cos^2 \varphi_l} \left(\frac{D_k}{24\Delta\lambda^2} + \operatorname{sgn}(p) \frac{M_k}{24\Delta\lambda} \right), \qquad p = \pm 2, \\ m_{k+p} &= \frac{1}{R^2 \cos^2 \varphi_l} \left(\frac{16 D_k}{24\Delta\lambda^2} + \operatorname{sgn}(p) \frac{8 M_k}{24\Delta\lambda} \right), \qquad p = \pm 1; \end{aligned} \quad (105)$$

— in the φ-direction we have equation (91), where

$$\begin{aligned} m_l &= \frac{30 D_l}{24 R^2 \Delta\varphi^2}, M_l = \frac{-D_{l+2} + 8D_{l+1} - 8D_{l-1} + D_{l-2}}{12\Delta\varphi} - D_l \tan\varphi_l, \\ m_{l+p} &= \frac{1}{R^2} \left(\frac{D_l}{24\Delta\varphi^2} + \operatorname{sgn}(p) \frac{M_l}{24\Delta\varphi} \right), \qquad p = \pm 2, \\ m_{l+p} &= \frac{1}{R^2} \left(\frac{16 D_l}{24\Delta\varphi^2} + \operatorname{sgn}(p) \frac{8 M_l}{24\Delta\varphi} \right), \qquad p = \pm 1. \end{aligned} \quad (106)$$

The properties of conservatism and dissipativity proved for the linear schemes (see Theorems 3, 4) also hold for the derived linearised schemes.

6. Numerical Experiments

The developed methods will now be tested using various linear advection-diffusion and several nonlinear diffusion problems. The linear problems are related to the global mass transport, where $\phi(\mathbf{x}, t)$ can be associated with the pollutant density, temperature, humidity, etc., in the Earth's atmosphere [4, 11], while the nonlinear models are typical in plasma physics, chemical kinetics, ecology, destruction of plasticity, etc. [8–10, 20, 23, 56, 64].

6.1. Linear Problems

6.1.1. Numerical Experiment 1 — Diffusion in a Spherical Sector

Let $\mathbf{U}(\mathbf{x}, t) \equiv 0$, $f(\mathbf{x}, t) \equiv 0$ and $\sigma \equiv 0$ in equation (1), while the diffusion coefficient $\mu = \text{Const} \neq 0$ only in a spherical sector. Then we intend to solve the linear diffusion problem

$$\phi_t = \nabla \cdot (\mu \nabla \phi), \qquad \phi(\mathbf{x}, 0) = \phi^0(\mathbf{x}),$$

where $\phi^0(\mathbf{x})$ is zero on the entire sphere S except for a local spot-like region. Figure 4 shows the initial condition and the solution (Method 1, with $\Delta\lambda = \Delta\vartheta = 1°$) at several time moments from $t = 20$ to $t = 100$ when the spot is at the equator. The diffusion of a spot initially located in a spherical sector near the North pole where $\mu \neq 0$ is presented in Figure 5 (Method 1). Figure 6 shows the diffusion process from the North pole using Method 2 (with $\Delta\lambda = \Delta\varphi = 6°$) with the diffusion coefficient of the form

$$\mu(\lambda, \varphi) \sim \sin^{p_1}\left(\frac{\lambda}{2}\right)^{p_2} \sin^2\varphi, \qquad p_1 = 6, \qquad p_2 = 2.$$

It is seen that the coefficient μ is significant mainly in three spherical sectors (Figure 6a).

From all the figures it follows that the initial spots are propagating as they are expected to: the diffusion process is developing only in those spherical sectors, where μ is nonzero or significant.

6.1.2. Numerical Experiment 2 — Diffusion Flux Over the Pole

Let $\mathbf{U}(\mathbf{x}, t) \equiv 0$, $\sigma(\mathbf{x}, t) \equiv 0$, while the diffusion coefficient $\mu = \text{Const}$ and $f(\mathbf{x}, t) = 10 \exp\{-75\rho^2(\mathbf{x}, \mathbf{x}_0)\}$ is the stationary forcing centered at $\mathbf{x}_0 = (\lambda_0, \vartheta_0) = (0.5°, 10°)$, where $\rho(\mathbf{x}) = \arccos(\mathbf{x} \cdot \mathbf{y})$ is the distance between two points of the unit sphere. At the initial moment the solution $\phi^0(\mathbf{x})$ is zero on the entire sphere (Figure 7a, $t = 0$). Over time, the spot is spreading uniformly in all directions and passes the North pole without any shape distortion (Figure 7a, $t = 10$ to 100; Method 1). The evinces that the discretisation of equation (1) around the pole cell leads to the numerical algorithm that correctly describes diffusion in the near pole area. In Figure 7b we plot the solution's 1D profile as well.

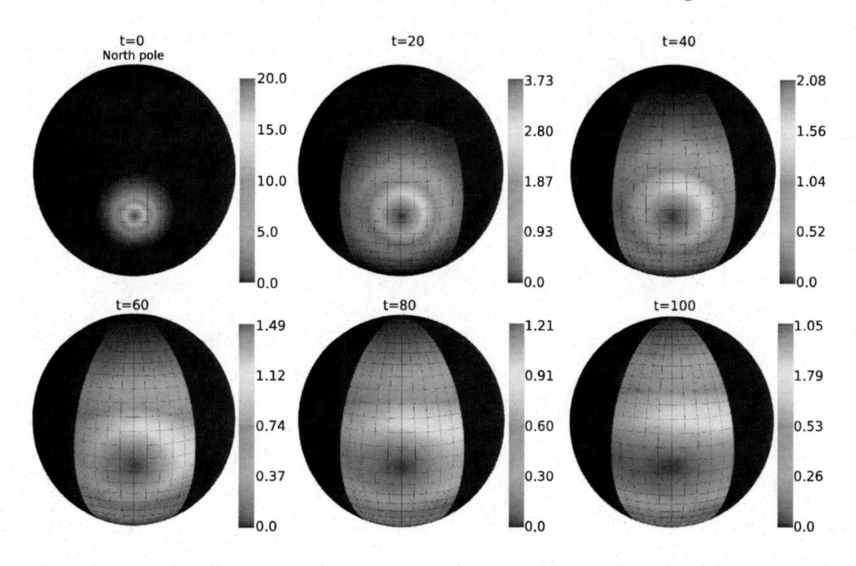

Figure 4. Diffusion in a spherical sector (Method 1): the initial condition at $t = 0$ and evolution of the solution $\phi^0(\mathbf{x}) \neq 0$ in a spherical sector from the equator.

6.1.3. Numerical Experiment 3 — Advection Flux Over the Pole

Let $f(\mathbf{x}, t) \equiv 0$, $\mu(\mathbf{x}, t) \equiv 0$, $\sigma(\mathbf{x}, t) \equiv 0$. Then equation (1) is reduced to the pure advection problem

$$\phi_t + \nabla \cdot (\mathbf{U}\phi) = 0, \qquad \phi(\mathbf{x}, 0) = \phi^0(\mathbf{x}).$$

Suppose that the velocity field $\mathbf{U}(\mathbf{x}, t)$ is directed over the poles (Figure 8a). Figure 8b shows the propagation of the initial condition, taken as a symmetric round spot (red-yellow), over the North pole. It is seen that the spot's propagation is nicely simulated: it is not distorted after passing over the pole. However, a remark to make is that the numerical scheme of Method 1 is not monotonic, since it is of the second approximation order with respect to the spatial variables [62]. In other words, since the phase velocity of a sinusoidal wave depends nonlinearly on its wave number, numerical dispersion appears that reduces the speed of propagation of short waves and violates the property of monotonicity for the scheme [40]. Because of this effect the Courant number

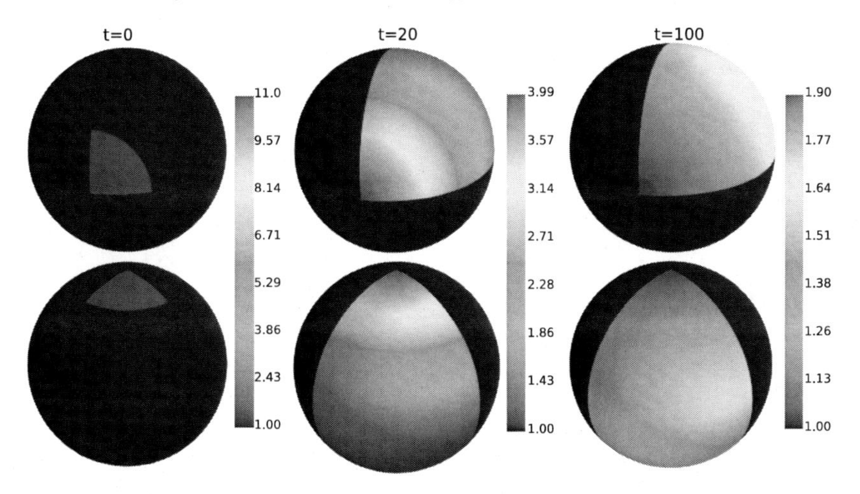

Figure 5. Diffusion in a spherical sector (Method 1): initial condition at $t = 0$ and evolution of the solution $\phi^0(\mathbf{x}) \neq 0$ in a spherical sector from the North pole (top row: view from the pole; bottom row: view from the equator).

$\tau \max\limits_{\mathbf{x}\in S} |\mathbf{U}(\mathbf{x},t)| / \min\{a\Delta\lambda \sin\vartheta, a\Delta\vartheta\}$ has to be small enough. And although, according to the Lax theorem, the dispersion decreases with τ, $\Delta\lambda$ and $\Delta\vartheta$, reducing the mesh sizes significantly increases the computation time. In our experiment we had $\Delta\lambda = \Delta\vartheta = 0.5°$, while the maximal value of the Courant number was 0.1.

6.1.4. Numerical Experiment 4 — Linear Advection-Diffusion Process

We now apply Method 1 to solve the linear advection-diffusion problem with the non-divergent wind velocity $\mathbf{U}(\mathbf{x},t)$ obtained from the ERA5 reanalysis data [63] (Figure 9a). Suppose that $\phi^0(\mathbf{x}) = 0$ and that the diffusion of a contaminant occurs with $\mu = \text{Const}$ and $\sigma(\mathbf{x},t) = 0$ from the following nine point sources

$$f(\mathbf{x},t) = \sum_{i=1}^{9} Q_i(\mathbf{x}_i,t), \qquad Q_i(\mathbf{x}_i,t) = \begin{cases} q_i = \text{Const}, & \text{if } \ 0 \le t \le 1 \text{ day} \\ q_i = 0, & \text{if } \ t > 1 \text{ day} \end{cases},$$

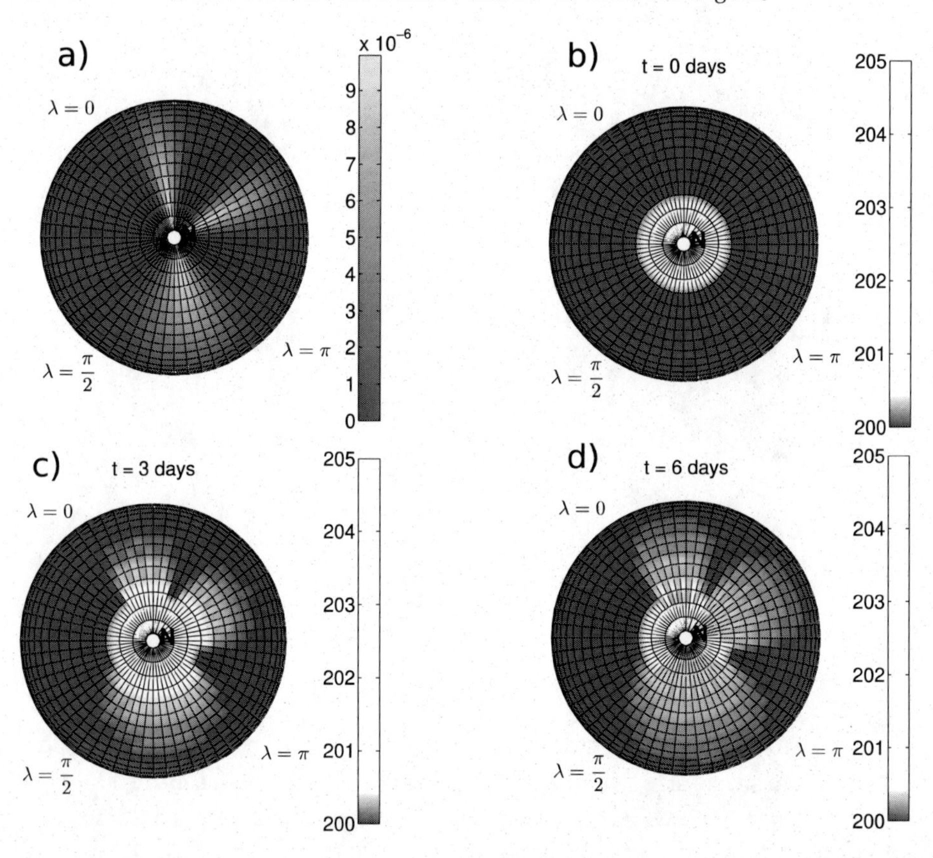

Figure 6. Diffusion in three sectors from the North pole (Method 2): (a) the profile of μ; (b) the initial condition at $t = 0$; (c) the solution at $t = 3$ days; (d) the solution at $t = 6$ days.

where q_i is the emission rate of the ith source. In this experiment we had $\Delta\lambda = \Delta\vartheta = 0.5^\circ$ and $\tau = 15$ sec. During the first day, the plumes of the contaminant propagate in the wind velocity direction, their areas increase due to the diffusion process, but they remain tied to the point sources. After one day, when the sources stop to emit, the plumes leave the sources and continue their propagation only under the influence of the velocity field and diffusion process (Figure 9b to d).

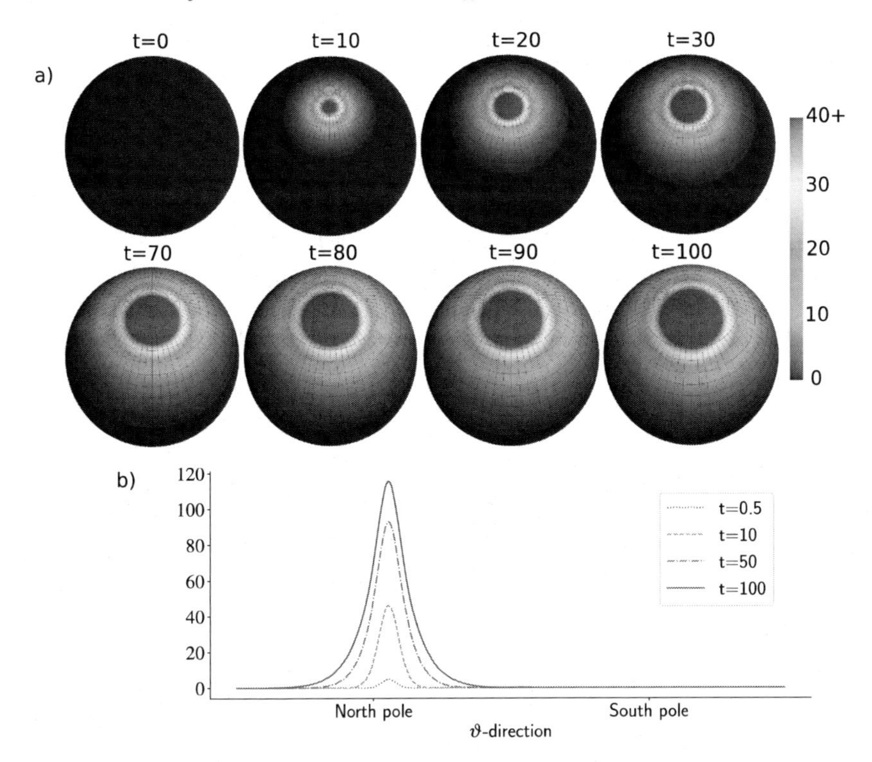

Figure 7. Diffusion of a spot near the pole (Method 1): (a) the initial condition and the solution at several time moments, evincing that the shape of the spot due to the forcing (the region inside the isoline $\phi(\mathbf{x}, t) = 40$ shown in red) keeps unchanged; (b) the solution's profile at different times.

6.2. Nonlinear Problems

Let $\mathbf{U}(\mathbf{x}, t) \equiv 0$, $\sigma(\mathbf{x}, t) \equiv 0$, $\mu = \mu(\phi)$ and $f = f(\phi)$. Then equation (1) describes a nonlinear diffusion process of the form

$$\phi_t = \nabla \cdot (\mu(\phi)\nabla\phi) + f(\phi), \qquad \phi(\mathbf{x}, 0) = \phi^0(\mathbf{x}). \tag{107}$$

In this case, the only change that needs to be made for Method 1 is the linearisation of system (28) in each double time interval (t_{n-1}, t_{n+1}), namely $\mu = \mu(\phi(t_{n-1}))$ and $f = f(\phi(t_{n-1}))$.

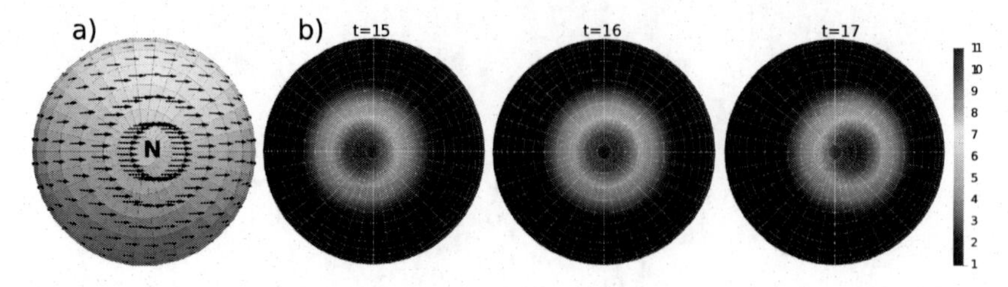

Figure 8. Advection of an initial spot over the North pole (Method 1): (a) the velocity field; (b) the solution at several time moments. The initial spot in shown in red and yellow, while the green and cyan is an effect of non-monotonicity of the scheme due to numerical dispersion.

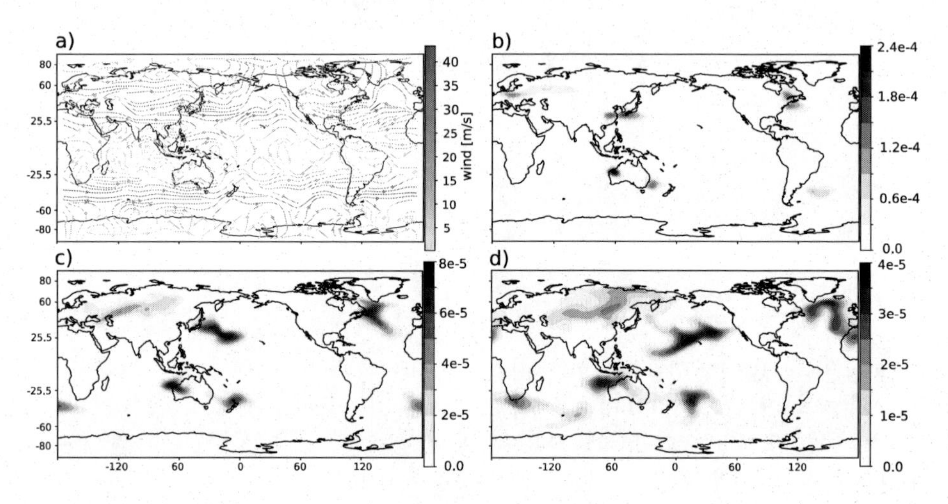

Figure 9. Advection-diffusion and non-stationary spatially distributed forcings (Method 1): (a) the non-divergent wind velocity; (b)-(d) the solution at several time moments.

6.2.1. Numerical Experiment 5 — Nonlinear Temperature Wave

If μ = Const and $f = \alpha\phi - \beta\phi^3$ then problem (107) describes a nonlinear combustion process at high temperatures. Figure 10 shows the numerical solution (Method 1) under the particular case of $\alpha = \beta$. It is seen that the initial burning region (the red spot) is expanding due to the homogeneous propagation of a nonlinear temperature wave of constant amplitude in all directions.

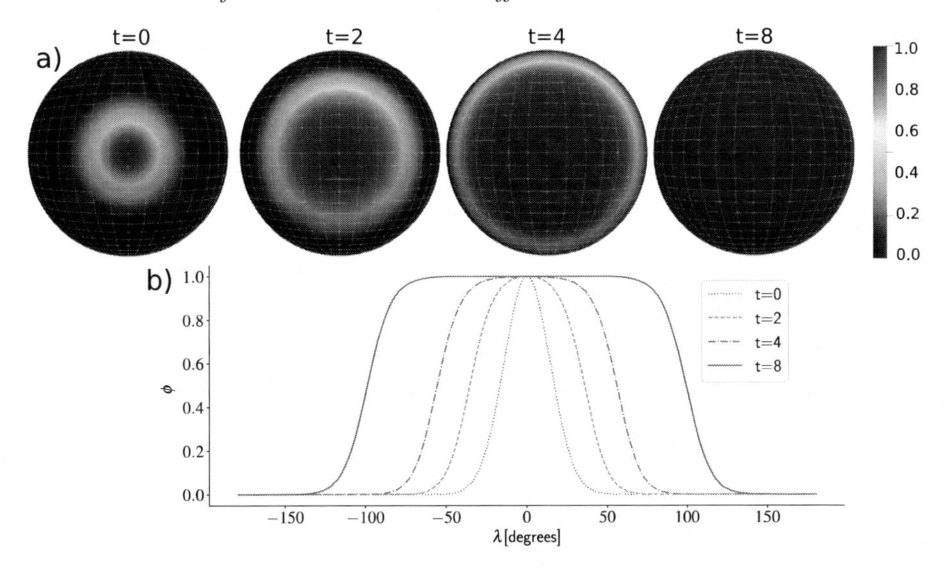

Figure 10. Nonlinear temperature wave (Method 1): (a) the solution at several time moments; (b) the temperature's profile in the λ-direction.

6.2.2. Numerical Experiment 6 — Blow-Up Nonlinear Combustion Regimes

Let $\mathbf{U} \equiv 0$, $\sigma \equiv 0$, $\mu = k\phi^{\alpha}$, $f = q\phi^{\beta}$, where $k, q > 0$. Then equation (107) describes nonlinear combustion processes, called it blow-up regimes, which are characterised by unlimitedly growing solutions within a finite time period. These modes appear due to a strong positive nonlinear feedback in the system: the greater the deviation from the equilibrium, the faster the process of combustion [2, 65]. It is worthy to remark that extremely growing solutions are widely encountered in real problems; they describe rapid compression and accumulation of matter in laser fusion, as well as a number of important processes in chemical kinetics, magnetohydrodynamics, meteorology (tornadoes and lightning), ecology (growth of biological populations), neurophysiology, epidemiology (infectious disease outbreaks), economics (rapid economic growth) and demography (world population growth) — just to name a few.

With Method 1, we have successfully simulated three blow-up combustion regimes: the HS mode, when the temperature rises rapidly in an expanding

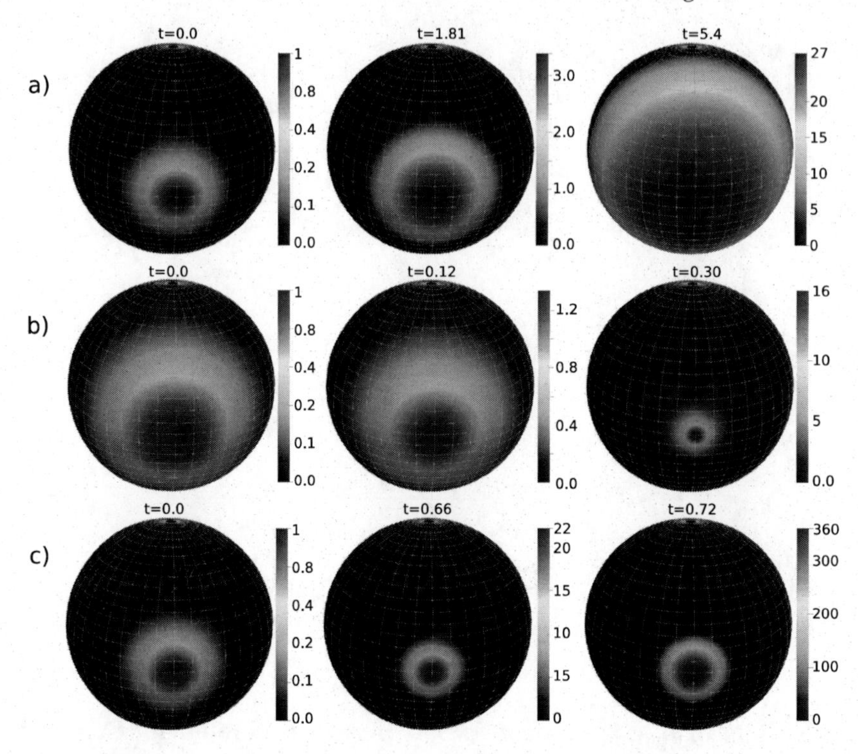

Figure 11. The regimes of combustion (Method 1): (a) HS mode ($\alpha = \beta = 1$); (b) LS mode ($\alpha = 1, \beta = 3$); (c) S mode ($\alpha = 1, \beta = 2$).

region ($\beta < \alpha + 1$, Figure 11a); the LS mode, when the temperature rapidly increases in a squeezing area ($\beta > \alpha+1$, Figure 11b); and the S mode, when the temperature rises rapidly in a region of constant size ($\beta = \alpha + 1$, Figure 11c).

6.2.3. Numerical Experiment 7 — Spiral Waves

The last experiment is aimed to test the schemes from two standpoints. First, we want to make sure that the schemes nicely work at various degrees on nonlinearity α; second, we have to study whether the solution to the split linearised problem (100)-(101) converges to the solution to the unsplit nonlinear problem (98). For this we calculate the forcing f of problem (98) whose analytical solu-

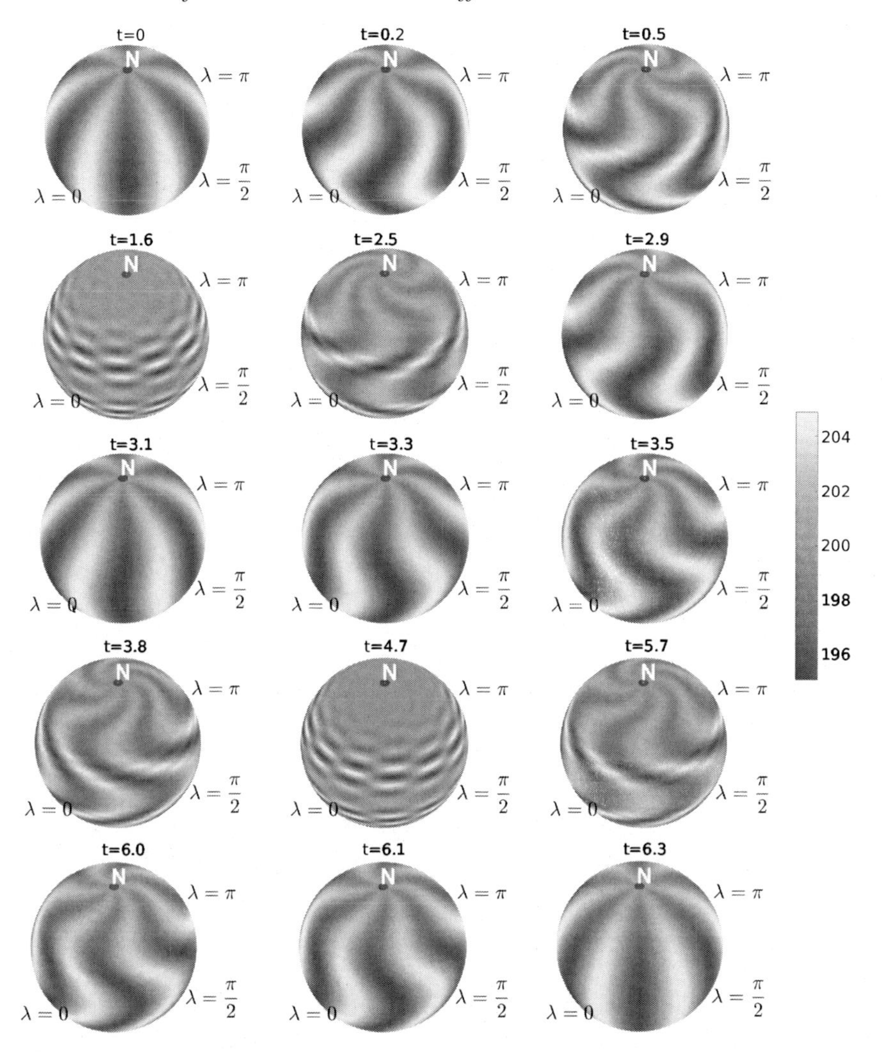

Figure 12. Spiral waves (Method 3): the numerical solution at several time moments.

tion is

$$T(\lambda, \varphi, t) = 5 \sin \xi \cos \varphi \cos^2 t + 200, \tag{108}$$

where

$$\xi = 6\lambda + 12\cos(4\varphi)\sin t, \tag{109}$$

under $\mu = 10^{-11}$. As one can see, formula (108) describes periodic oscillations of the solution in different directions (clockwise or anticlockwise), depending on the sign of $\cos(4\varphi)$.

The fourth-order numerical solution obtained with Method 3 is shown in Figure 12. The numerical solution is consistent with the analytics (108): according to the term $\sin\xi$ with $\cos(4\varphi)$, for $t \in (0, \pi)$ the spiral waves move in one direction, whereas when $t \in (\pi, 2\pi)$ then the direction is reversed. The entire cycle has the time period 2π. During this cycle, the spiral waves are completely destroyed several times. The method thus nicely reproduces the interaction of the three basic mechanisms — nonlinearity, external forcing and dissipation, that generate solutions with periodic breakups and renewals typical for complex structures and self-organisation.

CONCLUSION

Three balanced, implicit, unconditionally stable numerical methods are developed. All the methods essentially involve operator splitting as a basic numerical technique. One of the methods is a finite volume method of second approximation order both in space and time based on bordering the pole cells, which requires a prior determination of the solution at the poles. The other two methods are based on the grid-swap technique, which allows using periodic boundary conditions in both directions and thus constructing finite difference schemes of arbitrary (in particular, of second and fourth) approximation order in space and of second order in time. The numerical experiments with the three methods proved their ability to accurate simulating diverse linear and nonlinear advection-diffusion phenomena with constant and variable diffusion coefficients, including those that imply non-smooth solutions and high nonlinearity.

ACKNOWLEDGMENTS

The work was partially supported by the grant No. 14539 of the National System of Researchers of Mexico (SNI, CONACyT) and scholarship of CONACyT (Mexico).

REFERENCES

[1] Hundsdorfer, W. and Verwer, J. G., *Numerical solution of time-dependent advection-diffusion-reaction equations*, Springer Series in Computational Mathematics 33, Berlin Heidelberg, Springer, 2003.

[2] Kurdyumov, S. P., *Regimes with blow-up*, Moscow, Fizmatlit, 2006 (in Russian).

[3] Makarov, I. M. (ed.), *Blow-up regimes. Evolution of an idea: laws of co-evolution of complex structures*, Moscow, Nauka, 1999 (in Russian).

[4] Marchuk, G. I. and Skiba, Yu. N., Role of adjoint equations in estimating monthly mean air surface temperature anomalies, *Atmósfera* 5, 119–133 (1992).

[5] Nair, R. D. and Lauritzen, P. H., A class of deformational flow test cases for linear transport problems on the sphere, *J. Comput. Phys.* 229, 8868–8887 (2010).

[6] Pudykiewicz, J. A., Numerical solution of the reaction-advection-diffusion equation on the sphere, *J. Comput. Phys.* 213, 358-390 (2006).

[7] Skiba, Yu. N. and Filatov, D. M., On an efficient splitting-based method for solving the diffusion equation on a sphere, *Numer. Meth. Part. Diff. Eq.* 28, 331–352 (2012).

[8] Skiba, Yu. N. and Filatov, D. M., Numerical modelling of nonlinear diffusion phenomena on a sphere, in: N. Pina *et al.* (eds.), *Advances in Intelligent Systems and Computing* 197, Berlin Heidelberg, Springer, 2013, pp. 57–70.

[9] Skiba, Yu. N. and Filatov, D. M., An efficient numerical method for the solution of nonlinear diffusion equations on a sphere, in: A. R. Baswell (ed.), *Advances in Mathematics Research* 18, New York, Nova Science Publishers, 2013, pp. 271–298.

[10] Skiba, Yu. N. and Filatov, D. M., Modelling of combustion and diverse blow-up regimes in a spherical shell, in: P. Quintela *et al.* (eds.), *Progress in Industrial Mathematics at ECMI2016* 26, Berlin Heidelberg, Springer, 2017, pp. 729–735.

[11] Skiba, Yu. N. and Parra-Guevara, D., *Application of adjoint equations to problems of dispersion and control of pollutants*, New York, Nova Science Publishers, 2015.

[12] Zhang, Q., Johansen, H. and Colella, P., A fourth-order accurate finite-volume method with structured adaptive mesh refinement for solving the advection-diffusion equation, *SIAM J. Sci. Comput.* 34, B179–B201 (2012).

[13] Skiba, Yu. N., A non-iterative implicit algorithm for the solution of advectiondiffusion equation on a sphere, *Int. J. Numer. Meth. Fluids* 78, 257–282 (2015).

[14] Lacey, A. A., Ockendon, J. R. and Tayler, A. B., 'Waiting-time' solutions of a nonlinear diffusion equation, *SIAM J. Appl. Math.* 42, 1252–1264 (1982).

[15] Landsberg, C. and Voigt, A., A multigrid finite element method for reaction-diffusion systems on surfaces, *Comp. Vis. Sci.* 13, 177–185 (2010).

[16] Sherman, J. and Morrison, W. J., Adjustment of an inverse matrix corresponding to changes in the elements of a given column or a given row of the original matrix, *Ann. Math. Stat.* 20, 620–624 (1949).

[17] Sherman, J. and Morrison, W. J., Adjustment of an inverse matrix corresponding to a change in one element of a given matrix, *Ann. Math. Stat.* 21, 124–127 (1950).

[18] Thomas, L. H., Elliptic problems in linear difference equations over a network, *Watson Sci. Comput. Lab. Rept.*, New York, Columbia University, 1949.

[19] Lee, T. D., *Mathematical methods in physics*, New York, Columbia University, 1964.

[20] Bekman, I. N., *Mathematical theory of diffusion*, Moscow, Moscow University Press, 1990 (in Russian).

[21] Bear, J., Dynamics of fluids in porous media, New York, Dover Publications, 1988.

[22] Catté, F., Lions, P.-L., Morel, J.-M. and Coll, T., Image selective smoothing and edge detection by nonlinear diffusion, *SIAM J. Numer. Anal.* 29, 182–193 (1992).

[23] Glicksman, M. E., *Diffusion in solids: field theory, solid-state principles, and applications*, New York, John Wiley & Sons, 2000.

[24] Kametaka, Y., On the nonlinear diffusion equation of Kolmogorov-Petrovskii-Piskunov type, *Osaka J. Math.* 13, 11–66 (1976).

[25] King, J. R., 'Instantaneous source' solutions to a singular nonlinear diffusion equation, *J. Eng. Math.* 27, 31–72 (1993).

[26] Rudykh, G. A. and Semenov, E. I., Non-self-similar solutions of multidimensional nonlinear diffusion equations, *Math. Notes* 67, 200–206 (2000).

[27] Samarskii, A. A. *et al.*, *Blow-up in quasilinear parabolic equations*, New York, Walter de Gruyter, 1995.

[28] Vorob'yov, A. Kh., *Diffusion problems in chemical kinetics*, Moscow, Moscow University Press, 2003 (in Russian).

[29] Wu, Zh., Zhao, J., Yin, J. and Li, H., *Nonlinear diffusion equations*, Singapore, World Scientific Publishing, 2001.

[30] Frank-Kamenetskii, D. A., *Diffusion and heat exchange in chemical kinetics*, Moscow, Nauka, 1967.

[31] Peters, R., Peters, J., Tews, K. H. and Bähr, W., A microfluorimetric study of translational diffusion in erythrocyte membranes, *Bioch. et Biophys. Acta — Biomembranes* 367, 282–294 (1974).

[32] Koppel, D. E., Sheetz, M. P. and Schindler, M., Lateral diffusion in biological membranes. A normal-mode analysis of diffusion on a spherical surface, *Biophys. J.* 30, 187–192 (1980).

[33] Zykov, V. S. and Müller, S. C., Spiral waves on circular and spherical domains of excitable medium, *Phys. D: Nonlin. Phenomena* 97, 322–332 (1996).

[34] Yagisita, H., Mimura, M. and Yamada, M., Spiral wave behaviors in an excitable reaction-diffusion system on a sphere, *Phys. D: Nonlin. Phenomena* 124, 126–136 (1998).

[35] Tsynkov, S. V., Numerical solution of problems on unbounded domains. A review, *Appl. Numer. Math.* 27, 465–532 (1998).

[36] Skiba, Yu. N., Finite-difference mass and total energy conserving schemes for shallow water equations, *Russ. Meteorol. Hydrol.* 2, 35–43 (1995).

[37] Skiba, Yu. N. and Filatov, D. M., On splitting-based mass and total energy conserving arbitrary order shallow-water schemes, *Numer. Meth. Part. Diff. Eq.* 23, 534–552 (2007).

[38] Skiba, Yu. N. and Filatov, D. M., Conservative arbitrary order finite difference schemes for shallow-water flows, *J. Comput. Appl. Math.* 218, 579–591 (2008).

[39] Skiba, Yu. N. and Filatov, D. M., Mass and energy conserving fully discrete schemes for the shallow-water equations, in: G. Spadoni (ed.), *Energy Conservation: New Research*, New York, Nova Science Publishers, 2009, pp. 155–197.

[40] Marchuk, G. I., *Methods of numerical mathematics*, New York, Springer, 1982.

[41] Marchuk, G. I., *Splitting methods*, Moscow, Nauka, 1988 (in Russian).

[42] Lax, P. D. and Richtmyer, R. D., Survey of the stability of linear finite difference equations, *Comm. Pure Appl. Math.* 9, 267–293 (1956).

[43] Richtmyer, R. D., *Principles of advanced mathematical physics*, New York, Springer, 1978 (vol. 1) and 1981 (vol. 2).

[44] Baker Jr., G. A., An implicit numerical method for solving the n-dimensional heat equation, *Quart. Appl. Math.* 17, 440–443 (1960).

[45] Douglas Jr., J. and Rachford Jr., H. H., On the numerical solution of heat conduction problems in two and three space variables, *Trans. Amer. Math. Soc.* 82, 421–439 (1956).

[46] D'yakonov, E. G., Difference schemes of second-order accuracy with a splitting operator for parabolic equations without mixed derivatives, *Zh. Vychisl. Matem. i Matem. Fiz.* 4, 935–941 (1964) (in Russian).

[47] Godunov, S. K. and Ryaben'kii, V. S., *Finite difference schemes: An introduction to the theory*, Moscow, Nauka, 1977 (in Russian).

[48] Habetler, G. J. and Wachspress, E. L., Symmetric successive overrelaxation in solving diffusion difference equations, *Math. Comput.* 15, 356–362 (1961).

[49] Lobanov, A. I. and Petrov, I. B., *Numerical methods for solving partial differential equations (an online course)*, available at http://www.intuit.ru/department/calculate/nmdiffeq (2019).

[50] Özişik, M. N., *Finite difference methods in heat transfer*, CRC Press, 1994.

[51] Oliphant, T. A., An implicit numerical method for solving two-dimensional time-dependent diffusion problems, *Quart. Appl. Math.* 19, 221–229 (1961).

[52] Peaceman, D. W. and Rachford Jr., H. H., The numerical solution of parabolic and elliptic differential equations, *J. Soc. Industr. Appl. Math.* 3, 28–41 (1955).

[53] Samarskii, A. A., An economical difference method for the solution of the multidimensional parabolic equation in an arbitrary region, *USSR Comput. Math. Math. Phys.* 3, 894–926 (1963).

[54] Yanenko, N. N., *The method of fractional steps: Solution of problems of mathematical physics in several variables*, New York, Springer, 1971.

[55] Gibou, F. and Fedkiw, R., A fourth order accurate discretization for the Laplace and heat equations on arbitrary domains, with applications to the Stefan problem, *J. Comput. Phys.* 202, 577–601 (2005).

[56] Crank, J. and Nicolson, P., A practical method for numerical evaluation of solutions of partial differential equations of the heat-conduction type, *Math. Procs. Cambridge Phil. Soc.* 43, 50–67 (1947).

[57] Voevodin, V. V. and Kuznetsov, Yu. A., *Matrices and calculations*, Moscow, Nauka, 1984 (in Russian).

[58] Collatz, L., *Functional analysis and numerical mathematics*, Berlin, Springer, 1964 (in German).

[59] Strang, G., On the construction and comparison of difference schemes, *SIAM J. Numer. Anal.* 5, 506–517 (1968).

[60] Geiser, J., Fourth-order splitting methods for time-dependent differential equations, *Numer. Math.: Theory, Meth. Appl.* 1, 321–339 (2008).

[61] Press, W. H., Teukolsky, S. A., Vetterling, W. T. and Flannery, B. P., *Numerical recipes 3rd edition: The art of scientific computing*, Cambridge University Press, 2007.

[62] Godunov, S. K., A difference method for numerical calculation of discontinuous solutions of the equations of hydrodynamics, *Matem. Sbornik* 47, 271–306 (1959) (in Russian).

[63] European Centre for Medium-Range Weather Forecasts, The ERA5 dataset of hourly estimates of a large number of atmospheric, land and oceanic climate variables, available at https://www.ecmwf.int/en/forecasts/datasets/reanalysis-datasets/era5 (2019).

[64] Trusov, P. V. (ed.), *Introduction to mathematical modelling*, Moscow, Logos, 2004 (in Russian).

[65] Samarskii, A. A., Nonlinear effects of blow-up and localization processes in burning problems, in: C. M. Brauner and C. Schmidt-Laine (eds.), *Mathematical Modeling in Combustion and Related Topics*, Belgium, Martinus Nijhoff Publishers, 1988, pp. 217–231.

Contents of Earlier Volumes

Horizons in World Physics. Volume 301

Horizons in World Physics. Volume 300

Horizons in World Physics. Volume 299

Horizons in World Physics. Volume 298

Horizons in World Physics. Volume 297

INDEX

D

E

F

G

N

O

P

Q

R

S

T

V

W

X

Y